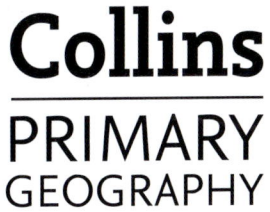

Collins PRIMARY GEOGRAPHY

Investigating
Pupil Book 3

Planet Earth	**Landscapes**	
	The Earth's surface	2
	The shape of the land	4
	Investigating landscapes	6
Water	**Water around us**	
	A wet planet	8
	The effects of water	10
	Looking down at water	12
Weather	**Weather worldwide**	
	Different types of weather	14
	Living in hot and cold places	16
	Sunshine matters	18
Settlements	**Villages**	
	A village community	20
	Different types of village	22
	Investigating villages	24
Work and Travel	**Travel**	
	Ways of travelling	26
	Finding your way	28
	Routes and journeys	30
Environment	**Caring for nature**	
	Wildlife around us	32
	Protecting wildlife	34
	Working together	36
Places	**Scotland**	38
	France	44
	South America	50
	Asia	56
	Glossary	62
	Index	63

Stephen Scoffham | Colin Bridge

Unit 1 Landscapes

Lesson 1: The Earth's surface

What is the Earth's surface like?

Key words
continent
Earth
landscape
planet
space
orbit

The Earth is one of eight planets that orbit around the sun. From far out in space the Earth looks quite small. You can see it is round like a ball.

The Earth is a very special planet. There is just the right mix of rock, air and water for animals and plants to live. No other planet is quite the same.

Water covers most of the Earth's surface. There are many different seas and oceans. The Pacific Ocean is the largest. The land is divided into great blocks called continents. Some of the land is very flat. In other places there are mountains, forests and deserts.

A

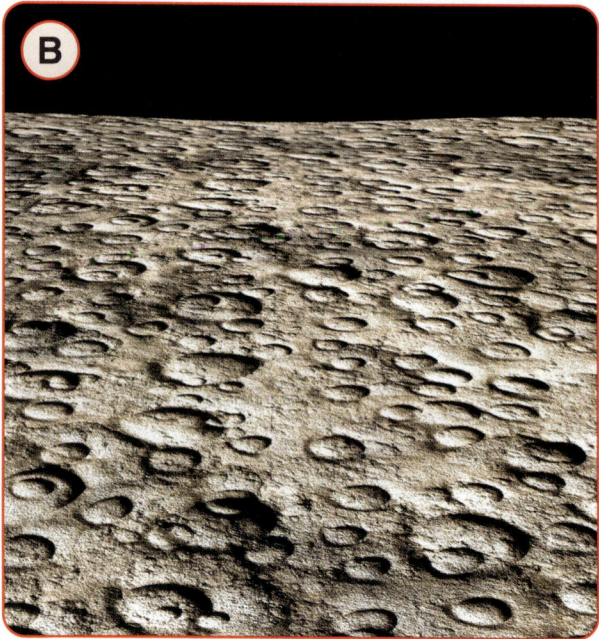

B

Discussion
- Which of the photographs on this page shows the Earth?
- What three clues helped you decide?
- Talk about how people might use the landscape shown in the photograph of the Earth.

| Unit 1 | Landscapes |

The Earth takes a year to orbit the sun.

Data bank
- The Earth is over 4000 million years old.
- Beneath the surface, the Earth is made of red hot rocks.
- At times in the past, the Earth was completely covered by ice.

Mapwork
Find a globe and turn it to the view in the space photograph. Name some of the places you see.

Investigation
Make a list of the colours you can see in the picture of the Earth on this page. Say what you think each one shows.

Unit 1 — Landscapes

Lesson 2: The shape of the land

Are all landscapes the same?

The shape of the land is called the landscape. Landscapes form very slowly over millions of years. Mountains are worn away by snow, ice, wind and rain. In other places, the land is rising. Geographers study how these changes happen.

Mountains are steep and rugged places. There is only a little soil for plants and the weather is often bad.

▼ The Himalayas in Asia.

Discussion
- Which landscape would be best for (a) rock climbing (b) walking?
- Why don't trees grow in every landscape?
- Which landscape is most similar to the place where you live?

Mapwork
Make a map or a model of an imaginary island with a number of different landscapes.

▲ A plateau in Chile, South America.

A plateau is a flat area found high up in the mountains. The weather here is often windy.

4

Unit 1 Landscapes

Hills are not as high as mountains but can have steep slopes made by rivers. There is enough soil for grass and trees to grow.

▼ Hills and valleys in France.

Islands are areas of land which are surrounded by water. They are often found in groups.

▼ An island in the Pacific Ocean.

Key words
coast
hill
island
landscape
lowland
mountain
plain
plateau
valley

Hills and valleys

Lowlands and plains

Coast

Islands

▲ Fields in Oxfordshire, southern England.

Lowlands and plains are flat landscapes. Many people live in lowland areas because they have the best farmland.

▲ The rocky coast of South Wales.

The coast is where the land meets the sea. Some coasts are rocky; others have sandy beaches, marshes or swamps.

Investigation
Start to make a geography notebook. Write down four landscape words and draw pictures to go with them.

Unit 1 Landscapes

Lesson 3: Investigating landscapes

What is the landscape like in Great Britain and Ireland?

Key words
Grampian Mountains
River Thames
Eryri (Snowdonia)
climate change

Great Britain and Ireland are made up of mountains, hills and lowlands. Most of the mountains are in the north and west. There are lowlands in the south and east. The rocks which make up the landscape date as far back as 700 million years.

Look at the map carefully. Do you live in Great Britain or Ireland or have you visited them? Which places do you know?

Data bank
- The Grampian Mountains have three or four months of snow each year.
- Climate change causes storms that can flood lowland areas.

Discussion
- Are Great Britain and Ireland mostly hilly or flat?
- Which colour on the map shows where most people live?
- Which sea is closest to where you live?

Mapwork
Working from the map, make a list of (a) mountain ranges (b) rivers.

6

Unit 1   Landscapes

A local enquiry

At St Mary's School the children did a project about their local landscape using a map and an aerial photograph. They had to imagine what it would look like if there were no buildings.

First the children listed all the landscape features such as hills, valleys and streams. Then they made a map of the area showing all the features on their list. Finally, they wrote a short report.

You could do a similar project about the place where you live.

▼ Aerial photograph of the local landscape.

Our school is close to the seashore. Some of the land is wet and marshy. The road out of town goes up a steep hill. As you go inland there are fields and farms.
In some places are woods.

Investigation

Find some pictures of different landscapes or cut them out from magazines. Write some sentences about each landscape for a class wall display.

Summary

In this unit you have learnt:

- about the surface of the Earth
- about different landscape features
- how to study the landscape.

7

Unit 2 Water around us

Lesson 1: A wet planet

Where do we find water?

Almost three-quarters of the Earth's surface is covered by water. Most of it is in the seas and oceans.

There is also a lot of water on the land. There is water in ponds, streams, rivers and lakes.

Wherever you live, water is all around you. It is in the rocks and soil under your feet. Water is also in the air and forms clouds above your head.

▼ The Victoria Falls are on the River Zambezi in Africa.

Key words
- cloud
- glacier
- iceberg
- lake
- pond
- river
- stream
- water vapour

Pie chart: Other water, Asia, Other land, Pacific Ocean

Discussion
- What are the different places where water is found?
- Where could you see, feel and hear water in your area?

Climate change
Find out what is happening to the ice in Greenland and Antarctica.

8

Unit 2 — Water around us

Clouds are made up of millions of tiny water droplets.

Liquid
When water droplets join together it starts to rain.

Gas
An invisible gas called water vapour fills the air around us.

Solid
If the air is very cold, the water droplets turn into snow or ice.

▲ Rain fills the dips and hollows in the land to make lakes.

▲ Snow can form into rivers of ice called glaciers.

▲ Rivers run down slopes to the sea.

▲ Icebergs are huge blocks of ice floating in the sea.

All water ends up in the sea.

Investigation
Make a drawing or plan of a house. Show in which rooms you would find (a) ice (b) running water (c) water vapour (steam).

Mapwork
Using an atlas make a list of lakes around the world.

Unit 2 Water around us

Lesson 2: The effects of water

Key words
- spider diagram
- pond
- soil
- water

Why is water important?

Without water everything would die. Trees and plants need water from the soil. People and animals use water for drinking and keeping themselves clean. Fish and many other animals need to live in water.

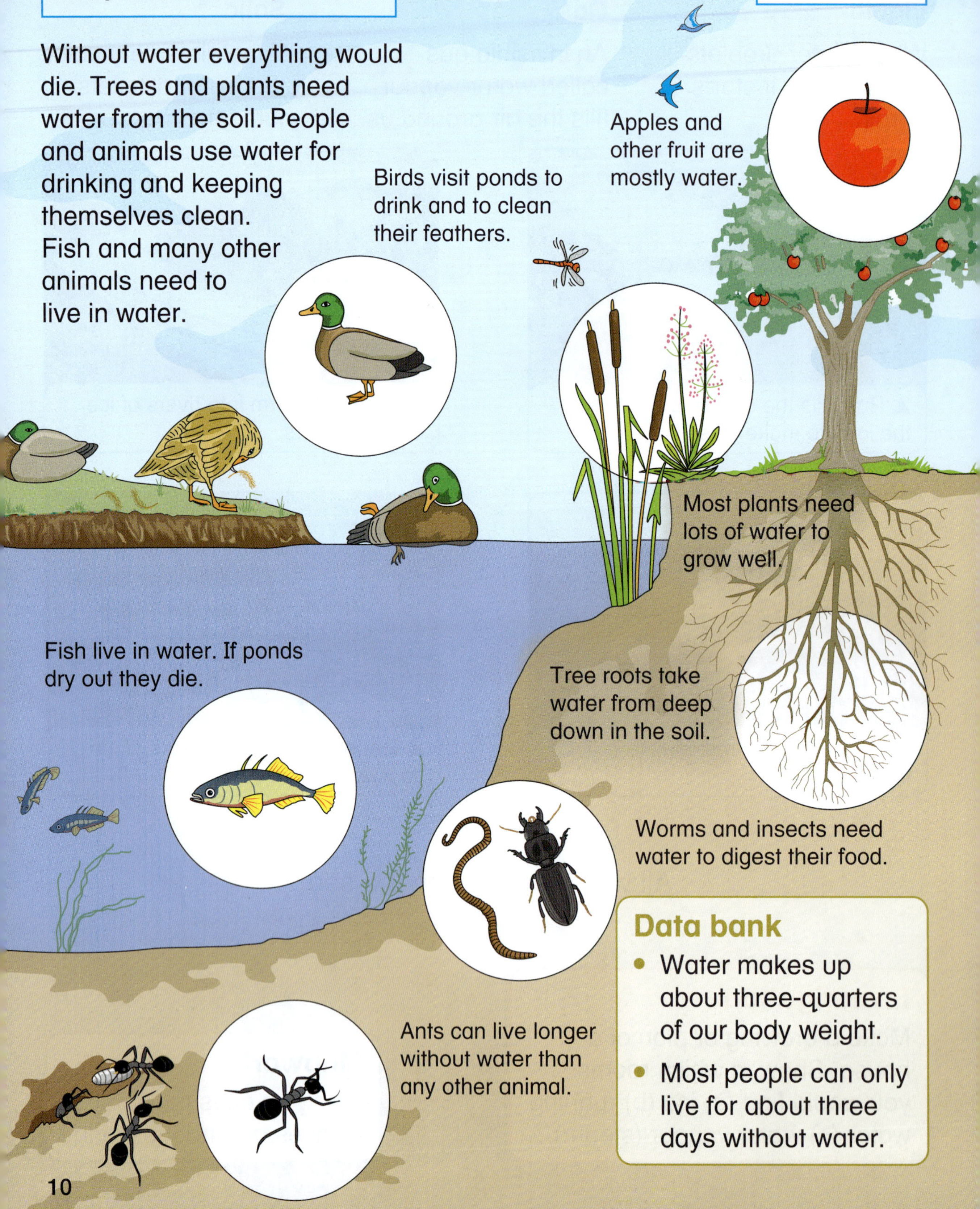

Birds visit ponds to drink and to clean their feathers.

Apples and other fruit are mostly water.

Most plants need lots of water to grow well.

Fish live in water. If ponds dry out they die.

Tree roots take water from deep down in the soil.

Worms and insects need water to digest their food.

Ants can live longer without water than any other animal.

Data bank
- Water makes up about three-quarters of our body weight.
- Most people can only live for about three days without water.

Unit 2   Water around us

Using water

▼ Drinking and cooking.

▼ Washing and cleaning.

▼ Watering plants.

▼ Travelling.

Discussion
- Why do living things need water?
- What are the main ways you use water?
- How do you think people, plants and creatures are affected by very dry weather?

Mapwork
Make a spider diagram. Write the word 'water' in the middle. Give examples of how it is used round the edge. Draw lines to complete the diagram.

Investigation
Look at the illustration of the pond. List plants and animals which (a) use water from the pond (b) use underground water.

11

Unit 2 — Water around us

Lesson 3: Looking down at water

> How is water shown in photographs from above?

Water shows up clearly in satellite images. It appears in rivers, marshes, reservoirs, lakes, seas and oceans. Water surrounds islands and marks the shape of the coast.

Key words
- satellite image
- marsh
- river
- coral island
- lake

Mapwork
Look at an atlas map of your country or continent. Make a list of any rivers, lakes and seas you can find.

▼ Marshes at the mouth of the Mississippi river.

▲ The twists and turns as a river flows to the sea.

▲ A coral island in the Pacific Ocean.

▼ The Great Lakes in the United States and Canada.

Unit 2   Water around us

Near our school

When rain reaches the ground it flows down hills and slopes. It can:

- run into streams and rivers
- stand in ponds and puddles
- soak into the soil
- go back into the air.

The roof, gutter, down pipes and drainpipes stop the rain from coming into buildings. Paint protects wood from rotting and metal from rusting.

Investigation

Make a survey of where rainwater goes in your school grounds. Write a report or draw a picture with notes around the edge.

Discussion

- Which are the wettest and driest places around your school?
- How does water disappear from puddles, ponds and lakes?
- Can you find any place names that involve water from a map of your area?

Summary

In this unit you have learnt:

- about the different forms of water, where water is found, and where it goes
- why water is important to animals, plants and people
- how water is shown on maps.

13

Unit 3 — Weather worldwide

Lesson 1: Different types of weather

Is the weather the same all over the world?

There are different types of weather around the world. Some places are very hot. Others are cold. In some places it rains a lot. In others it is very dry. The pattern of weather over a number of years is called the climate.

Key words
climate
desert
North Pole
polar lands
rainforest

Data bank
- The Sahara Desert is over 5000 km across.
- The North Pole and Antarctica are covered by vast sheets of ice.
- The Amazon rainforest is nearly the size of Europe.

Discussion
- Compare the climate in the desert, rainforest and polar lands.
- What would you put in a survival kit for each place?
- What animals might live in each region? Which has the hardest job to survive and which the easiest?

▲ **Rainforest**
In the rainforest the climate is hot and wet. Huge numbers of plants and animals live in the trees. Wide rivers flow through the forest.

| Unit 3 | Weather worldwide |

▲ **Desert**

In the desert the climate is very dry. Plants and animals have to live on very little water. Some deserts are very hot.

▲ **Polar lands**

Polar lands have the coldest climate on Earth. In some places snow and ice last all year. Very few plants and animals can live there.

Mapwork

Add drawings of the rainforest, desert and polar lands to your geography notebook. Write a sentence about each one.

Climate change

Research online to find out about extreme weather events and climate change around the world.

15

Unit 3 Weather worldwide

Lesson 2: Living in hot and cold places

How do people live in hot and cold places?

Key words
date palm
desert
market
oasis
polar lands
temperature

In the past, people found it hard to survive. They had to grow their own food, make their own clothes and keep their own animals. They used whatever materials they could find to build houses. If the weather was too hot or too cold their crops or animals could die.

Today, people can live almost anywhere in the world. Electricity, machines and modern homes help us to cope with difficult weather. Food can be delivered by lorry or by plane. However the weather still affects us.

Discussion
- How do different types of weather affect your life?
- What are the main problems of living in desert and polar lands?
- What should you do on very hot days?

Living in polar lands

In Greenland some people earn a living from fishing. What do you think is happening in these two photographs?

Unit 3 Weather worldwide

Living in the desert

There are only a few places in the desert where people can find water. These places are called oases. Some oases are small villages. Others are towns and have important markets.

▼ An oasis in Morocco.

Many desert houses have flat roofs because there is very little rain.

Camels carry heavy loads but they are being replaced by cars and trucks.

Food crops, like wheat, grow well in the desert when they have enough water.

All life depends on water in the desert.

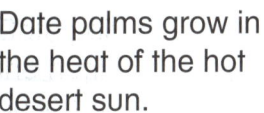
Date palms grow in the heat of the hot desert sun.

Mapwork
Make a picture map or a model of an oasis.

Investigation
Draw two things in a hot desert and two things from polar lands that are different to where you live.

Unit 3 Weather worldwide

Lesson 3: Sunshine matters

Why are some places hot and other places cold?

Key words
equator
North Pole
South Pole
hibernate

At the equator the sun rises high in the sky. The sun's rays fall straight on to the Earth and heat it up.

Near the North and South poles, the sun is always low in the sky. There are long shadows and the air never gets warm.

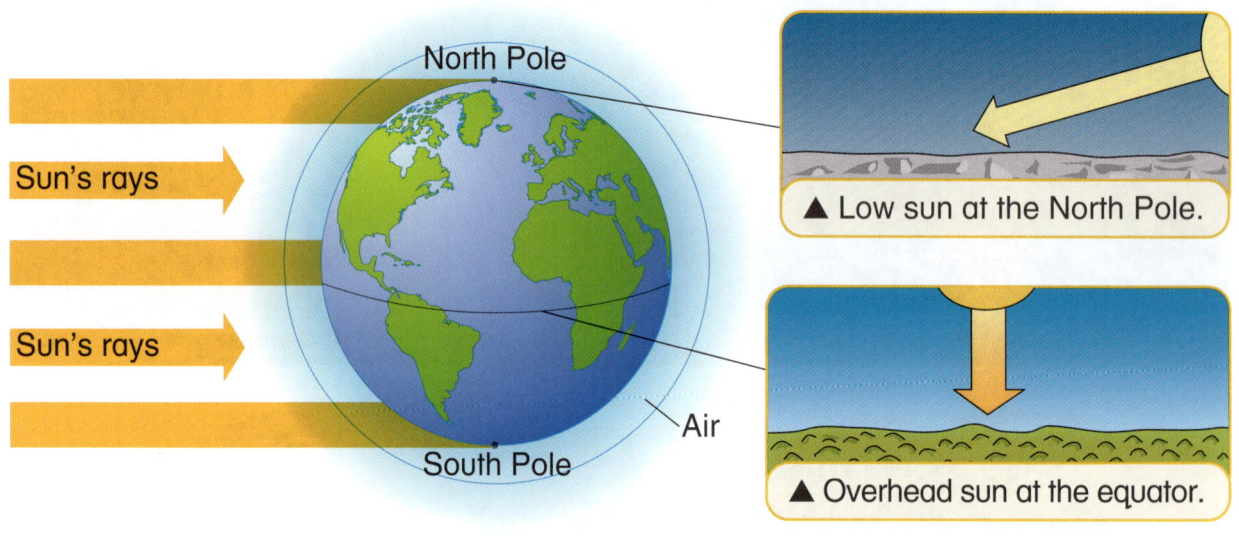

▲ Low sun at the North Pole.

▲ Overhead sun at the equator.

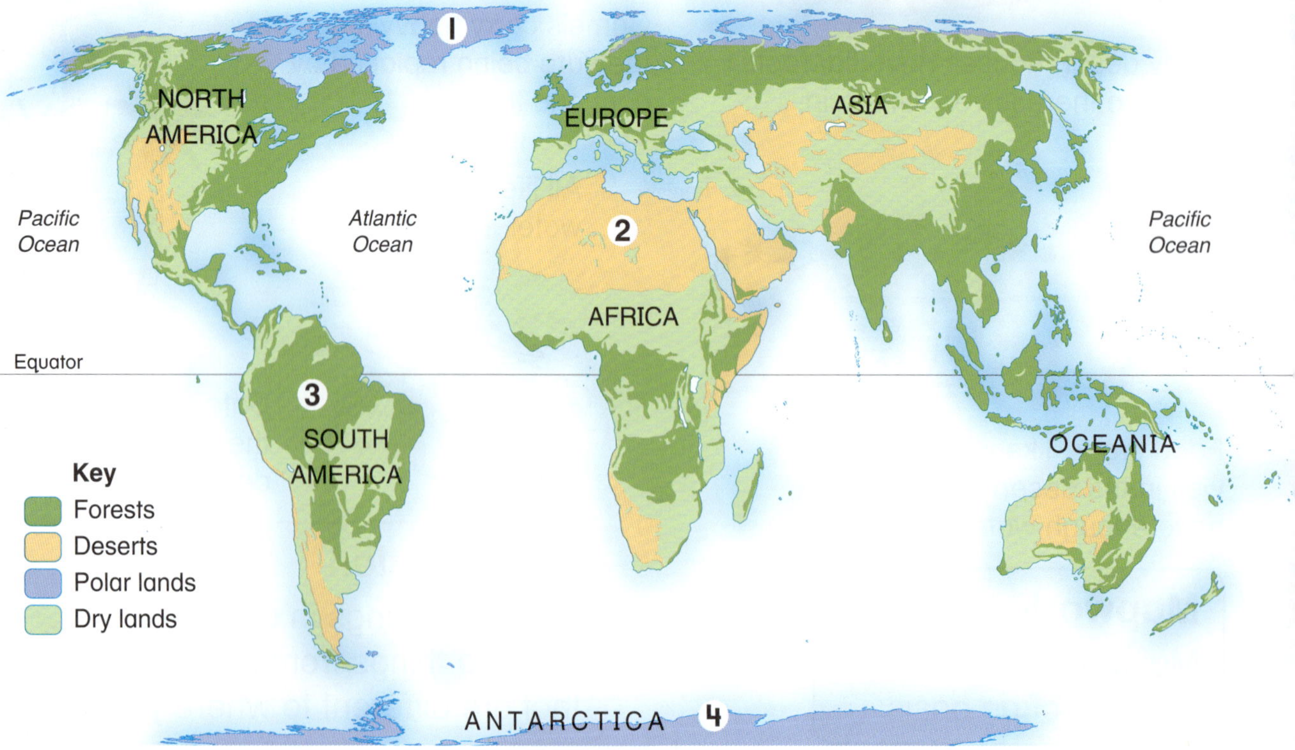

Key
- Forests
- Deserts
- Polar lands
- Dry lands

18

Unit 3 Weather worldwide

Hot and cold

Some countries have hot and cold weather at different times of the year. The UK and other parts of Western Europe can be hot in the summer and cold in the winter.

In the summer the ground dries out and the soil becomes dusty. People like to go to the beach to stay cool. In the winter there can be snow. Ponds and lakes freeze over. Plants stop growing and animals hibernate.

▼ A dry lake in France.

▼ Deep snow in Germany.

Discussion
- How would you describe the weather where you live?
- Why are there no deserts or ice caps in the UK?

Mapwork
Match the numbers from the map to the places in this list.

Sahara Desert • Antarctica • Amazon rainforest • Greenland

Investigation
Where are the hardest climate conditions for plants and creatures in your school grounds?

Summary
In this unit you have learnt:
- that some parts of the world are always hot and other parts are always cold
- how to describe the weather in words and diagrams
- how the weather affects animals, plants and people.

Unit 4 Villages

Lesson 1: A village community

What makes a village?

A village is a place where a group of people have made their homes close together. If people start living on other planets they will have to build places where they can live safely. They will need shelter, food and water. In the past, villages were built here on Earth for the same reasons.

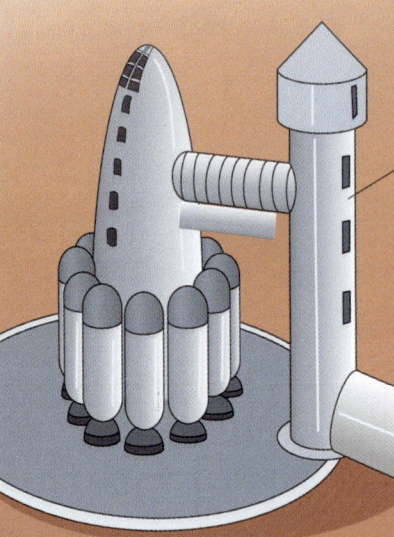

▲ Land use map for a settlement on Mars.

Key words

community
land use
planet
settlement
transport
village

Discussion

- Which areas in your town or village match the areas in the Mars village?
- Which do you think are the most important areas of the Mars village? Explain why.

20

Unit 4 | Villages

Mapwork
Make a land use map for different areas in your school.

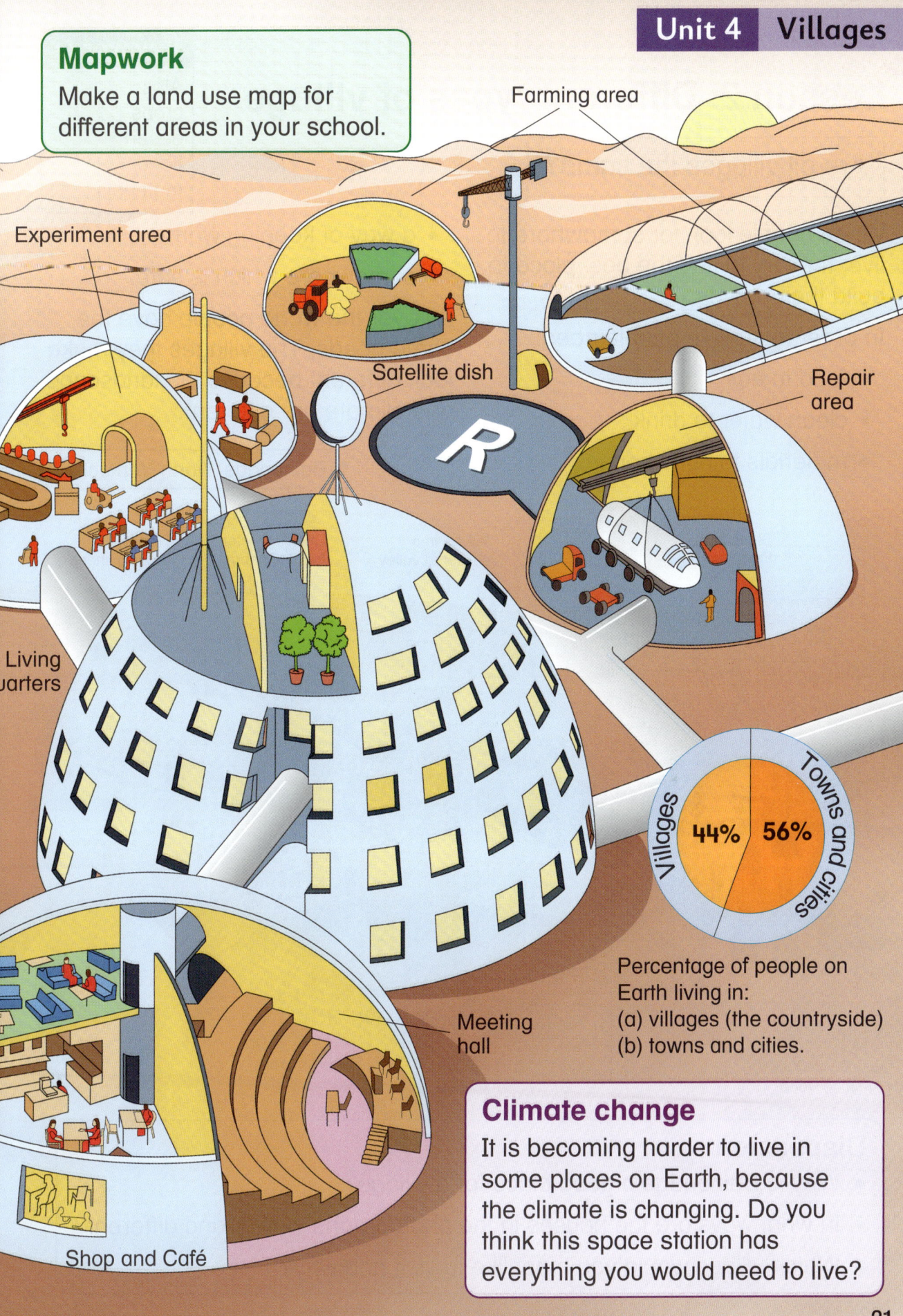

Farming area
Experiment area
Satellite dish
Repair area
Living quarters
Meeting hall
Shop and Café

Villages 44% | Towns and cities 56%

Percentage of people on Earth living in:
(a) villages (the countryside)
(b) towns and cities.

Climate change
It is becoming harder to live in some places on Earth, because the climate is changing. Do you think this space station has everything you would need to live?

21

Unit 4 Villages

Lesson 2: Different types of village

Are all villages the same?

Key words
Alps
crops
desert
flood
materials

When people look for somewhere to live, they try to find the best place to build their homes.

In order to survive people need:
- food to eat
- clean water to drink
- materials to build houses
- a way of keeping warm
- somewhere which is safe.

All over the world people have the same needs. The villages they make look different because the landscape and climate are different.

▼ Bainbridge, North Yorkshire.

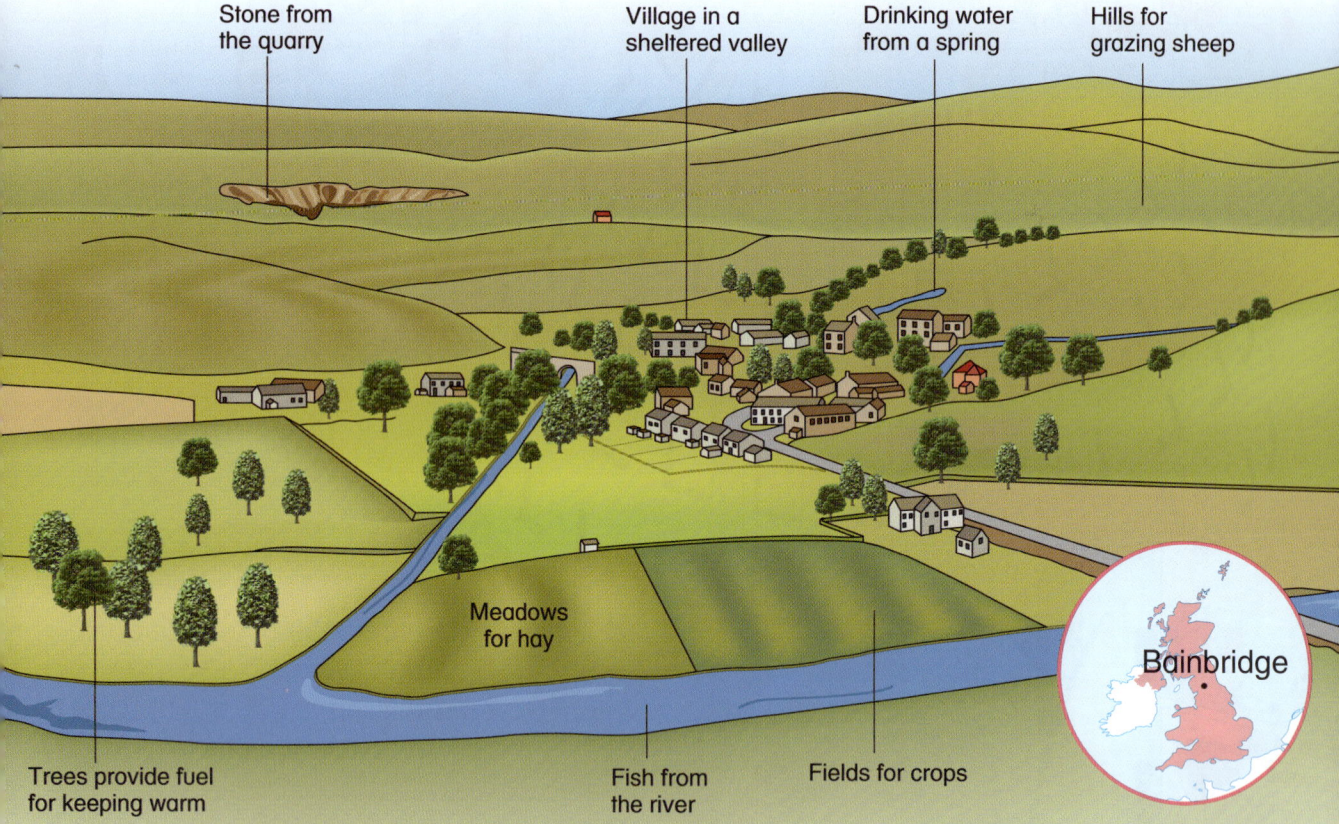

Stone from the quarry
Village in a sheltered valley
Drinking water from a spring
Hills for grazing sheep
Trees provide fuel for keeping warm
Meadows for hay
Fish from the river
Fields for crops
Bainbridge

Discussion
- Why is Bainbridge a good place for a village?
- In what ways are the houses in the photographs similar and different?
- Which village would you most like to visit and why?

Unit 4 Villages

▶ **A desert village**
Burkina Faso, West Africa

This village is on the edge of the Sahara Desert. The houses give shade and shelter from the sun. They have been grouped together for safety. The walls are made from baked earth and the roofs from dried grass. People keep goats and cattle.

◀ **A lake village**
Myanmar, Southeast Asia

The people in this village make their living by fishing. They eat some of the fish and sell the rest in the market. The wood for their houses comes from the forest nearby. The houses are built on stilts to keep the village safe from floods.

▶ **A mountain village**
Switzerland, Central Europe

This village is high in the Alps. The winters are cold and it often snows. The houses are made of stone and wood. Their roofs hang down over the walls to shed the snow. Some people in the village keep cows. The cows feed in the high meadows above the village in the summer.

Mapwork
Working from a local map, make a list of villages in your area or region.

Investigation
Draw pictures of the three types of houses described and display them around a world map to show their location.

23

Unit 4 Villages

Lesson 3: Investigating villages

How do villages change?

Key words
bungalow
church
greenhouse
orchard
paddock
farm

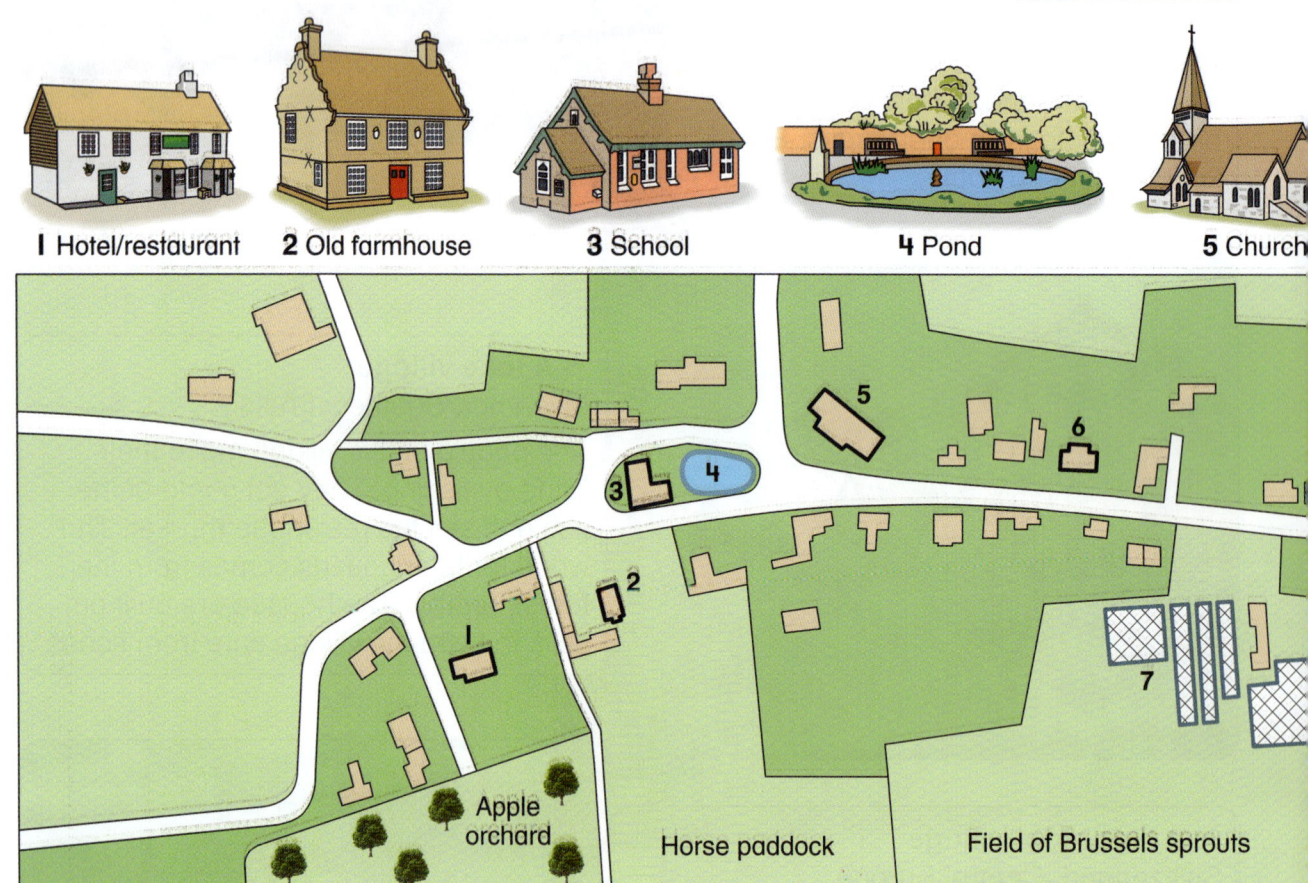

1 Hotel/restaurant 2 Old farmhouse 3 School 4 Pond 5 Church

Worth is a village in southeast England. Two hundred years ago there was just a group of buildings around a church. In Victorian times new houses were added and a school was built.

Worth now has a playing field and many more houses. A farmer has built some large greenhouses for tomatoes. The village is bigger than before but still has farmland all around it.

Many English villages are like Worth. They have grown larger as people who work in the towns move to the country.

Discussion
- What are the oldest parts of a village?
- How has Worth changed over the last two hundred years?
- What new feature would you add to Worth?

24

Unit 4 Villages

Data bank
- Most of the villages in England were founded before the Norman invasion of 1066.
- Villages vary in size from less than a hundred people to several thousand or more.
- Some people like to live in a village but travel by car to work elsewhere.

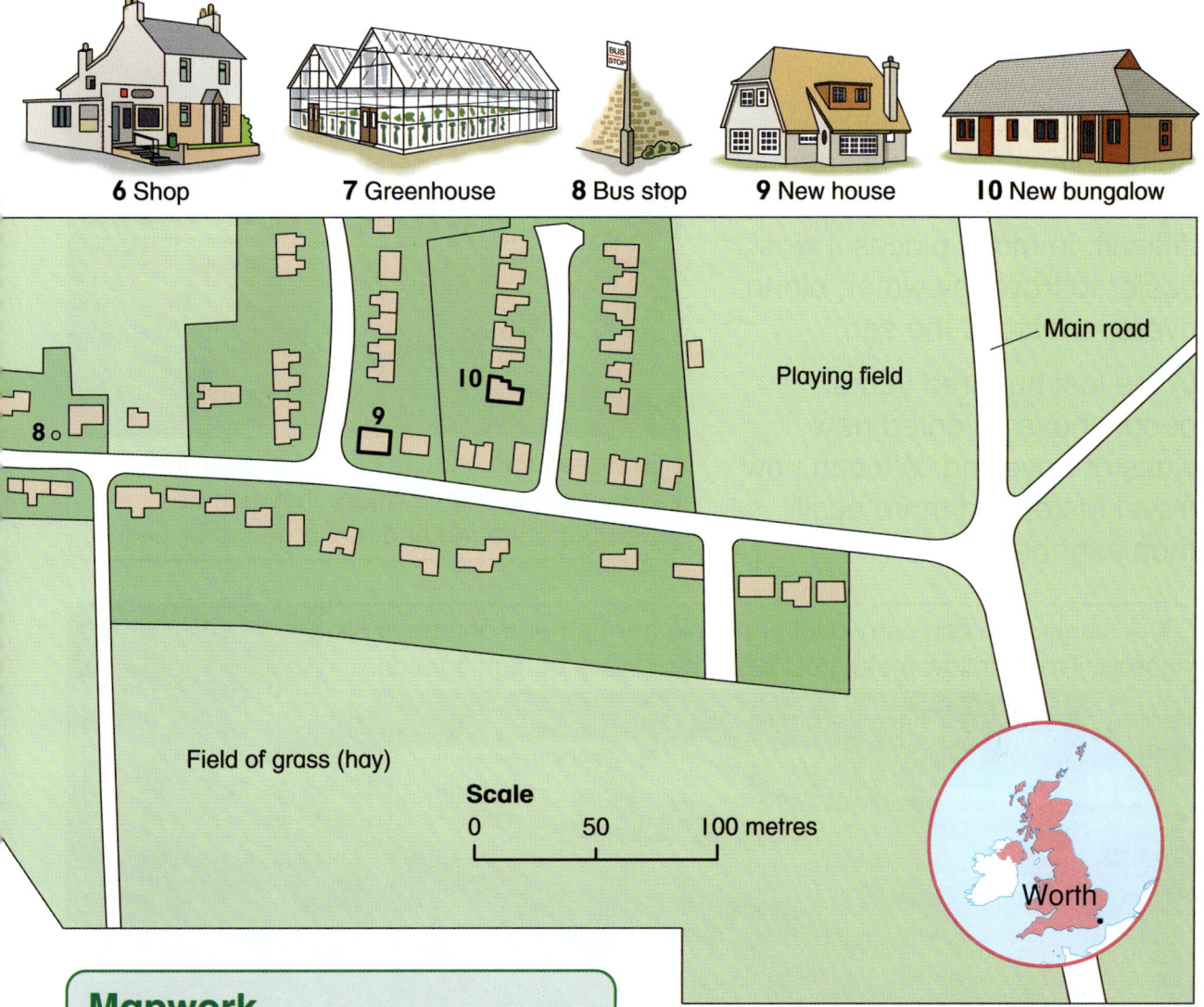

6 Shop **7** Greenhouse **8** Bus stop **9** New house **10** New bungalow

Mapwork
Draw pictures and plans of five buildings in Worth.

Investigation
Explore a village for yourself. Record what you find in photographs, maps and drawings.

Summary
In this unit you have learnt:
- what a village is
- about different villages around the world
- how to study a village.

Unit 5 Travel

Lesson 1: Ways of travelling

What different types of transport are there?

In the past, horses and carts helped people to move from place to place. Sometimes thick forests and high mountains made travel difficult. In many places it was easier to travel by water, along rivers, or across the sea.

In the last hundred years, people have invented new ways of travelling. We can now travel faster and more easily than ever before.

▲ Buses and coaches travel along roads. They need garages where they can be cleaned, repaired and recharged or refuelled.

▼ A single train can carry a lot of people or hundreds of tonnes of goods. Trains follow metal tracks and can travel at high speeds.

Discussion
- What are the main ways of travelling?
- How has transport changed?
- Which vehicle would you like to drive and why?

Key words
airport
harbour
transport
vehicle
goods

Unit 5 Travel

◀ Ships and boats take people, vehicles and goods across water. They need harbours or sheltered places where they can tie up and unload.

▼ Aeroplanes are the fastest way of travelling. They can cross mountains and seas without difficulty. They need airports for taking off and landing.

Data bank
- Around half of primary school children in the UK walk to school.
- There are 41 million vehicles in the UK.
- Nearly 200 million people use UK airports each year.
- There are more cars in China than in any other country in the world.

Climate change
All vehicles cause pollution and contribute to climate change. Talk about the problems they cause and what can be done about them.

Investigation
Make a survey of how children in your class (a) travel to school (b) travel about at weekends.

Unit 5 Travel

Lesson 2: Finding your way

Why do people use maps?

When people travel to a place for the first time, they need to think about which way to go. A map helps them to plan their route.

There are many types of map. Some show the centre of a town. They are drawn to a large scale and show a lot of detail. Others mark railway lines. They are drawn to a smaller scale and are not as detailed.

People who are driving, walking or travelling by train look for the map which shows their route best.

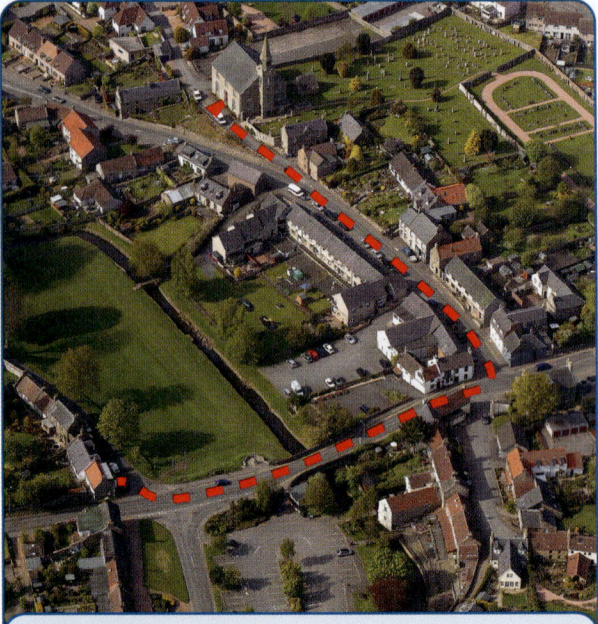

▲ Route from a house on the main road to the local church.

Discussion
- Why do people need different types of map?
- When do you use a map?

▲ **Van driver**

I deliver parcels to schools. I use a satnav because it shows street names.

▲ **Walker**

When I am planning a walking trip, I use a map which marks the hills, footpaths, bridges and villages.

▲ **Planner**

When I plan new roads and houses, I draw maps to show different designs.

Unit 5 Travel

Key words

grid
route
scale
Ordnance Survey

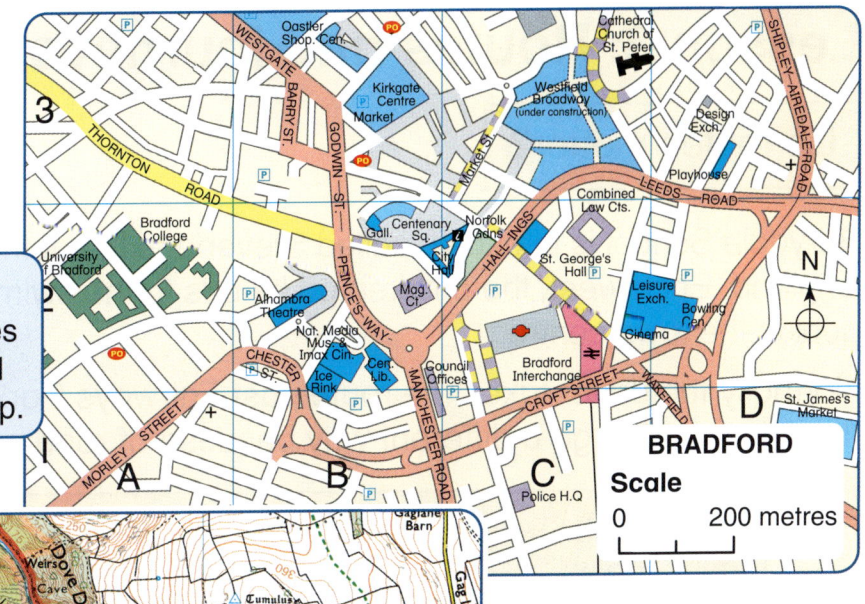

▶ **Street maps**

A grid of small squares makes it easier to find places on a street map.

◀ **Footpath maps**

Many footpath maps are made by Ordnance Survey. They show buildings and the shape of the land.

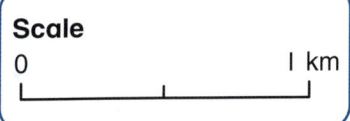

◀ **Train maps**

Train maps show rail routes between places.

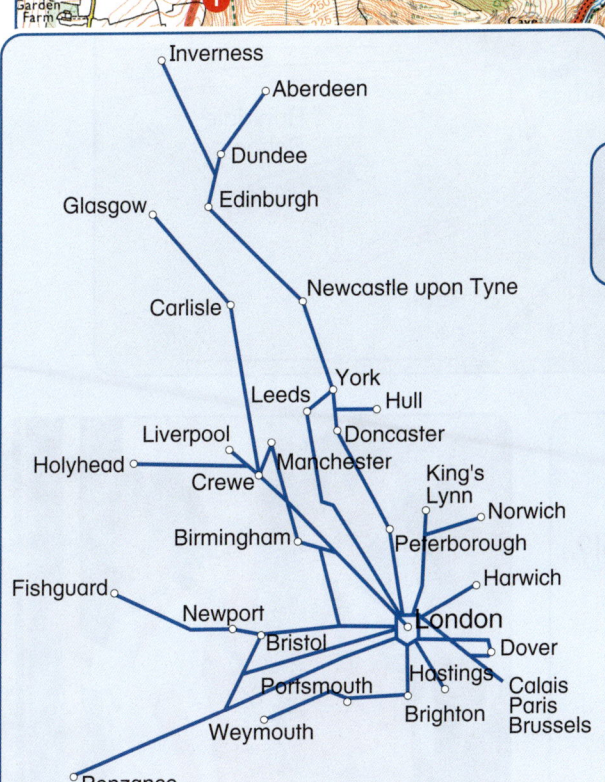

Mapwork

Draw a map to show a visitor how to reach your class from the school entrance.

Investigation

Make a class display of different maps of your area, your country or wider world.

Unit 5 | Travel

Lesson 3: Routes and journeys

Do routes matter?

Key words
journey
landmark
Ordnance Survey
routes

Burydale School is in a town called Stevenage in southern England. Each week the children go by bus to the swimming pool in the town centre.

Normally the driver takes the route past Fairways Park. One week the driver had to go a different way because of road works.

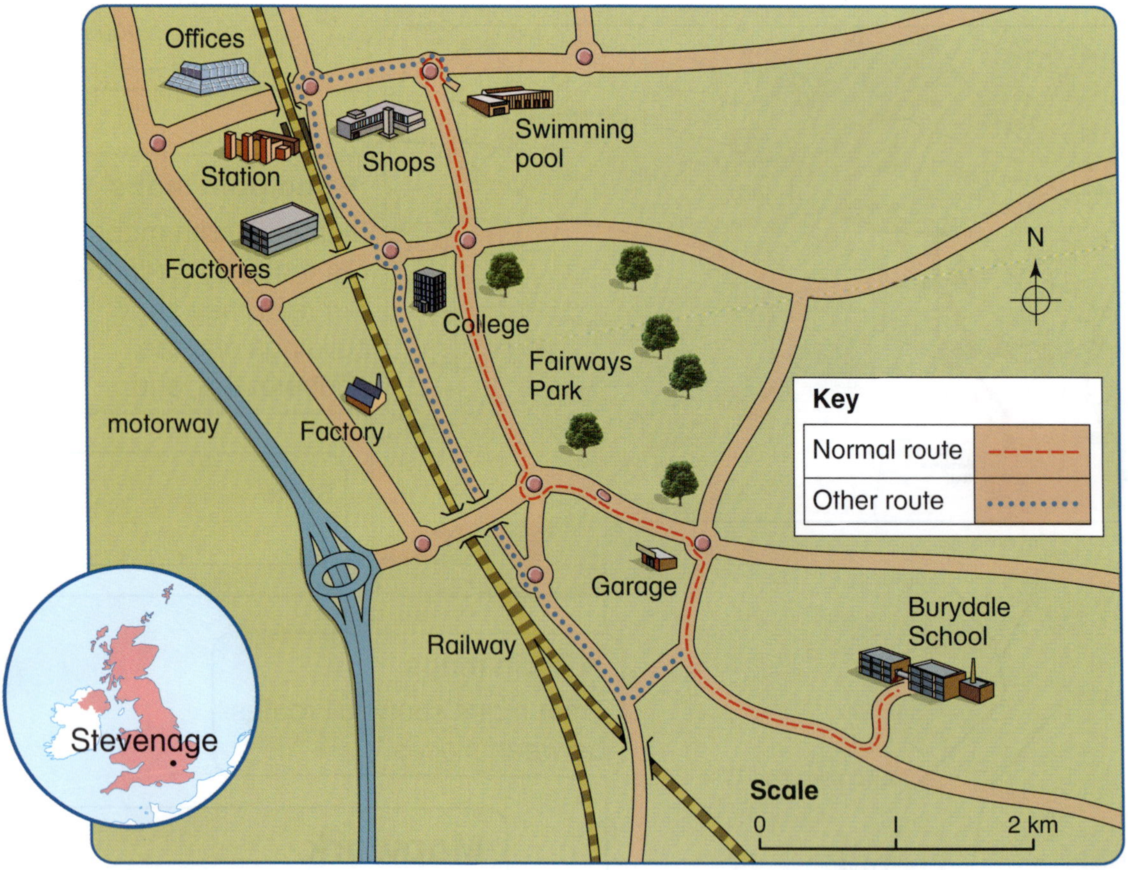

Key
Normal route ‒ ‒ ‒ ‒
Other route ......

Scale
0   1   2 km

Discussion
- What landmarks did they pass on the new route to the swimming pool?
- What landmarks are shown on the child's map opposite?
- Why do you sometimes take different routes to the same place?

30

Unit 5 Travel

As part of their project on routes, the children talked about their journeys from home to school.

They made a list of the landmarks which they passed and then drew maps of their journeys.

Data bank
- The oldest maps in the world are cave paintings made thousands of years ago.
- Modern maps use computers to help store and sort data.
- In the UK, Ordnance Survey maps were first made for the army to use in times of war.

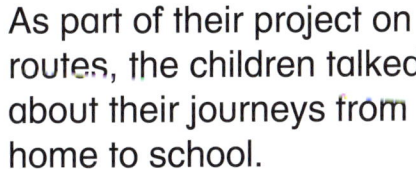

Mapwork
Make a list of the landmarks you pass on your way from home to school. Show them on your own route map.

Investigation
Display the results for how children in your class travel to school on a bar chart.

Summary
In this unit you have learnt:
- what a route is
- how people use maps for journeys
- why landmarks are important.

Unit 6 — Caring for nature

Lesson 1: Wildlife around us

What is a habitat?

The place where a community of plants and animals live is called a habitat. Ponds, woods, hedges, fields, waste ground and old walls are examples of different habitats. They provide food, water and shelter for the many plants and animals which live there.

Looking for a home
- butterfly
- worm
- woodlouse
- blackbird
- spider
- snail

Discussion
- Where might all the creatures that are looking for a home find a place to live in the old wall?
- What makes the old wall a good nature habitat?
- Can you think of any parts of your school grounds which are a bit wild?

Data bank
- As their numbers drop, people are trying to make safe places for creatures such as otters, water voles, beavers and red squirrels.
- There are ten national parks in England, three in Wales and two in Scotland.

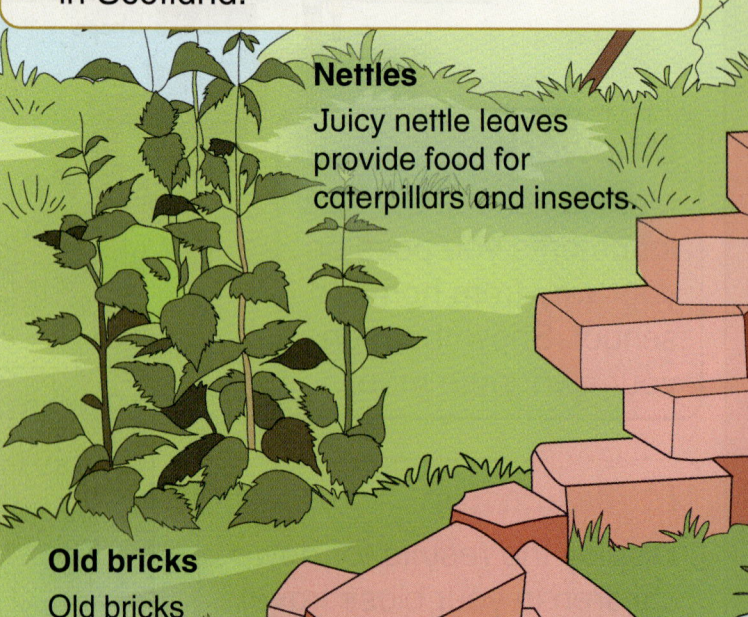

Nettles
Juicy nettle leaves provide food for caterpillars and insects.

Old bricks
Old bricks provide shelter for small animals.

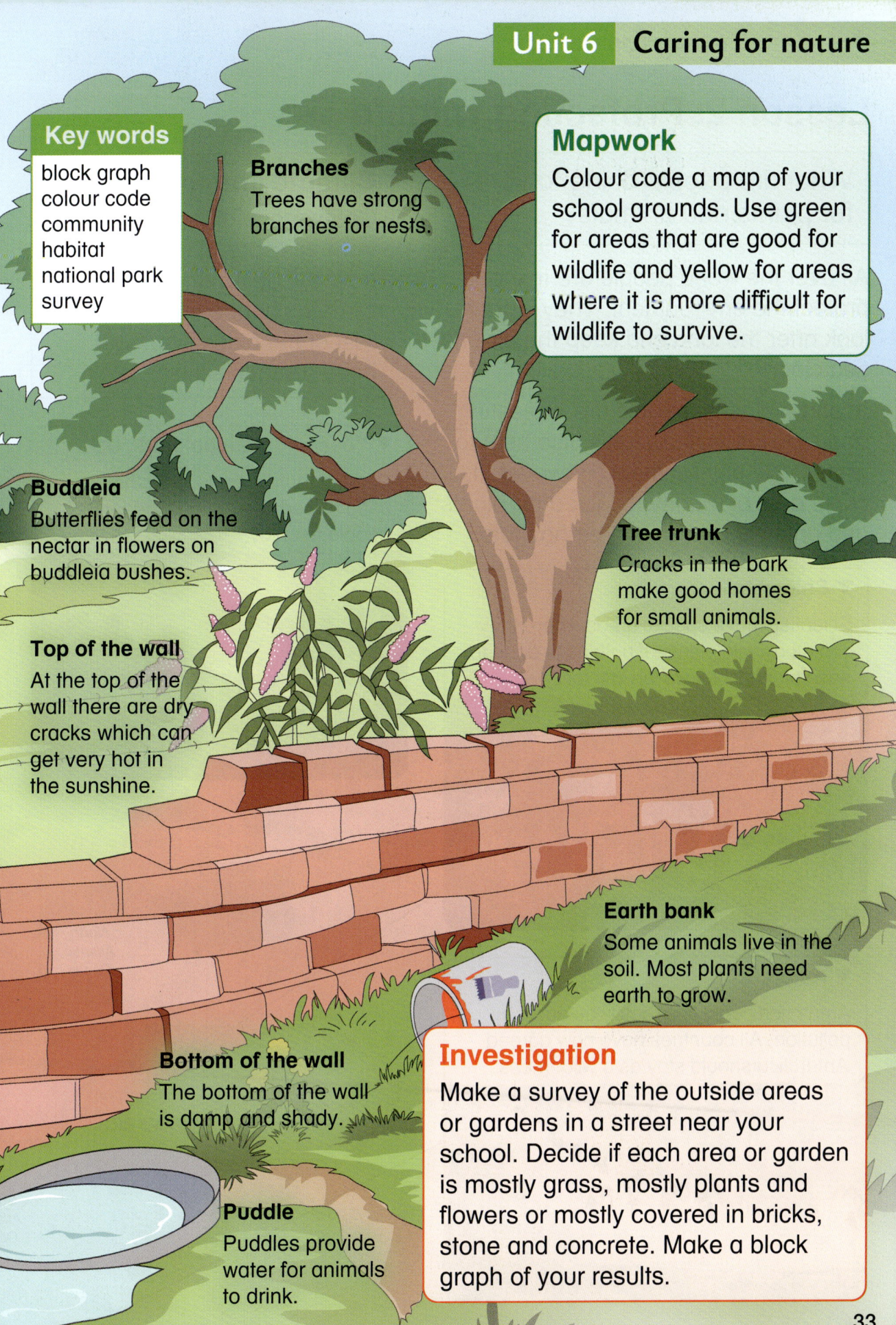

Unit 6 Caring for nature

Key words
block graph
colour code
community
habitat
national park
survey

Branches
Trees have strong branches for nests.

Mapwork
Colour code a map of your school grounds. Use green for areas that are good for wildlife and yellow for areas where it is more difficult for wildlife to survive.

Buddleia
Butterflies feed on the nectar in flowers on buddleia bushes.

Tree trunk
Cracks in the bark make good homes for small animals.

Top of the wall
At the top of the wall there are dry cracks which can get very hot in the sunshine.

Earth bank
Some animals live in the soil. Most plants need earth to grow.

Bottom of the wall
The bottom of the wall is damp and shady.

Investigation
Make a survey of the outside areas or gardens in a street near your school. Decide if each area or garden is mostly grass, mostly plants and flowers or mostly covered in bricks, stone and concrete. Make a block graph of your results.

Puddle
Puddles provide water for animals to drink.

Unit 6 Caring for nature

Lesson 2: Protecting wildlife

> What are people doing to care for plants and animals?

All over the world people are trying to protect the environment. They want to look after the land and keep the air and sea clean.

In some places, people are protesting about pollution. New laws also help to protect the environment. However, it costs a lot of money to look after plants and animals and save their habitats.

Climate change
Choose one of the places from this spread and research how it is being affected by climate change. Tell the rest of the class what you have discovered.

▼ **Yosemite National Park, US**
Some beautiful landscapes have been turned into national parks for people to enjoy.

▼ **Colombia**
Scientists have special areas where they can study wild plants. If nothing is done to save plants, they will be lost for ever.

▼ **Antarctica**
Antarctica could very easily be spoilt by pollution. All countries have now agreed Antarctica should stay as a wilderness.

34

Unit 6 Caring for nature

Discussion
- Why is it important to protect the environment around the world?
- Which conservation project do you think is most important?
- What could you do to protect wildlife?

Key words
conservation	national park
environment	pollution
habitats	reserve

Mapwork
Make a map of your nearest national park or conservation area.

Investigation
Explore ways to encourage people to care for wildlife.

▼ Kenya
Game reserves have been set up to protect lions, elephants and other animals.

▲ China
In China, thousands of trees have been planted to stop the desert from spreading over farmland.

▲ Southern Ocean
People are trying to save whales from hunting which could make them extinct.

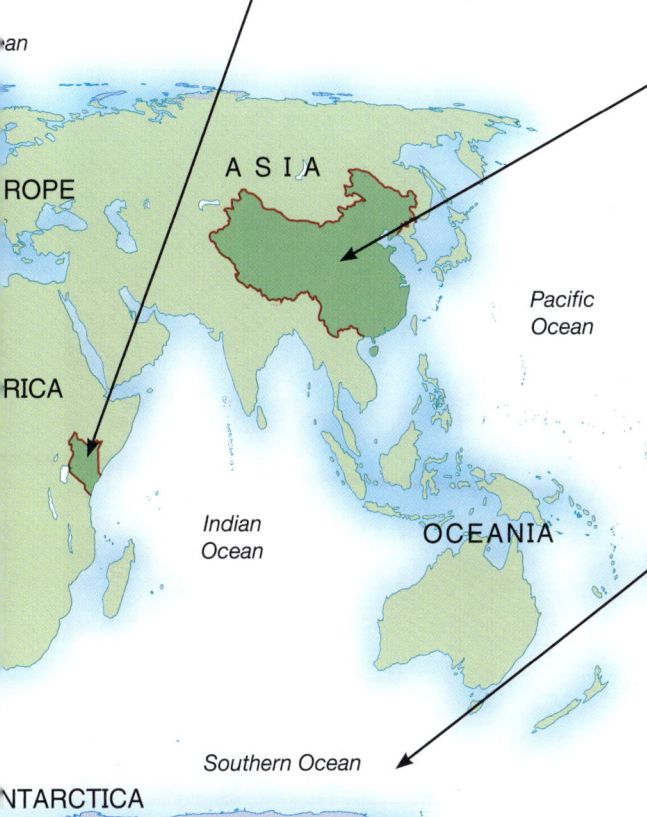

35

Unit 6 Caring for nature

Lesson 3: Working together

What happens at a nature reserve?

Key words
coral   reef
pollution   warden

The Great Barrier Reef in Australia is a world famous nature reserve. It is home to thousands of fish, plants and shell creatures. Warmer seas and pollution mean that it is now in danger. People are working together to protect it for the future:

- Scientists are studying the plants and animals that live there.
- Local people are helping with information about it.
- Wardens are looking after some of the islands.
- Volunteers are cleaning up rubbish and pollution.
- The government has passed laws to ban fishing.

▼ The Great Barrier Reef in Australia.

Data bank
- The Great Barrier reef is over 2000 kilometres long.
- The reef can be seen from outer space.
- It is the world's largest structure made by living creatures.

▼ Cleaning plastic from the Great Barrier Reef.

Unit 6   Caring for nature

Improving the school grounds

At Byron School the children joined a nature club. The club helped them to make their school grounds a better habitat for plants and animals.

Climate change

Gardens with flowers and trees attract birds and insects, keeping them healthy. This improves their chance of surviving extreme weather.

▲ The children designed and made a vegetable garden.

▼ The children put food out to attract the birds.

▲ The garden is used for school project work.

Mapwork

Draw a plan of your school garden, or design a garden plan of your own.

Investigation

Devise a short nature trail for your school grounds.

Summary

In this unit you have learnt about:

- what makes a habitat
- how people are caring for habitats around the world.

37

Unit 7   Scotland

Lesson 1: Introducing Scotland

What is Scotland like?

Landscape
Many parts of Scotland have high mountains. There are lowlands in the middle of Scotland. The west coast has lots of islands.

Weather
In the Scottish mountains it is often wet and cold. During the winter, snow can cover the ground for many months at a time.

◀ I live in Fort William. It rains on over 250 days in the year.

◀ Aberdeen has less than half the rainfall of Fort William.

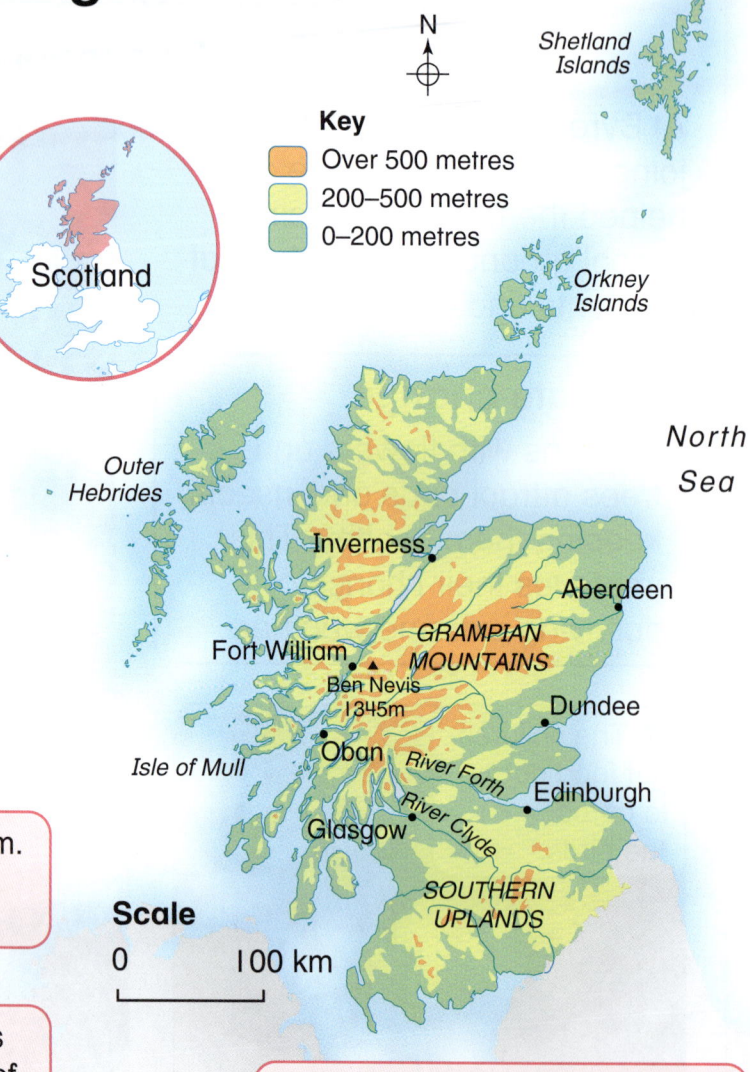

Key
- Over 500 metres
- 200–500 metres
- 0–200 metres

Shetland Islands
Orkney Islands
Outer Hebrides
Inverness
Aberdeen
Fort William
GRAMPIAN MOUNTAINS
Ben Nevis 1345m
Dundee
Isle of Mull
Oban
River Forth
River Clyde
Edinburgh
Glasgow
SOUTHERN UPLANDS
North Sea

Scale 0 — 100 km

▲ Scotland is the most northerly part of the United Kingdom.

▲ Autumn in the Scottish mountains.

38

Unit 7 Scotland

Key words
United Kingdom
Ben Nevis
Edinburgh
Glasgow
River Clyde
Shetland Islands
forestry

Mapwork
List the islands shown on the map. Add others to your list using an atlas.

Discussion
- What three things would you tell a visitor about Scotland?
- How is Scotland different to where you live?
- Where do you think you would most like to live in Scotland?

Transport

The lowlands and east coast have the main road and rail routes. Ferries deliver food and goods to the islands.

Settlement

Edinburgh and Glasgow are the largest cities. Aberdeen and Dundee are two ports on the east coast. Very few people live in the mountains.

▲ Playing bagpipes at Glasgow Green.

Work

Many people work in factories, tourism, and the oil industry. Farming, fishing and forestry are also important. Scotland is famous for tartan cloth and bagpipes.

▲ A lot of the electricity used in Scotland comes from wind farms.

Investigation
Research online to find out more about (a) tourism (b) wind farms in Scotland.

Unit 7   Scotland

Lesson 2: Edinburgh: The capital city of Scotland

What is Edinburgh like?

Edinburgh is a very old city. A castle was built there hundreds of years ago. It stands on a high rocky crag above the city.

Today Edinburgh is a busy place with lots of traffic. People come a long way to go shopping in the city. Tourists arrive by plane, train and car. Lots of people work in the shops, offices and factories.

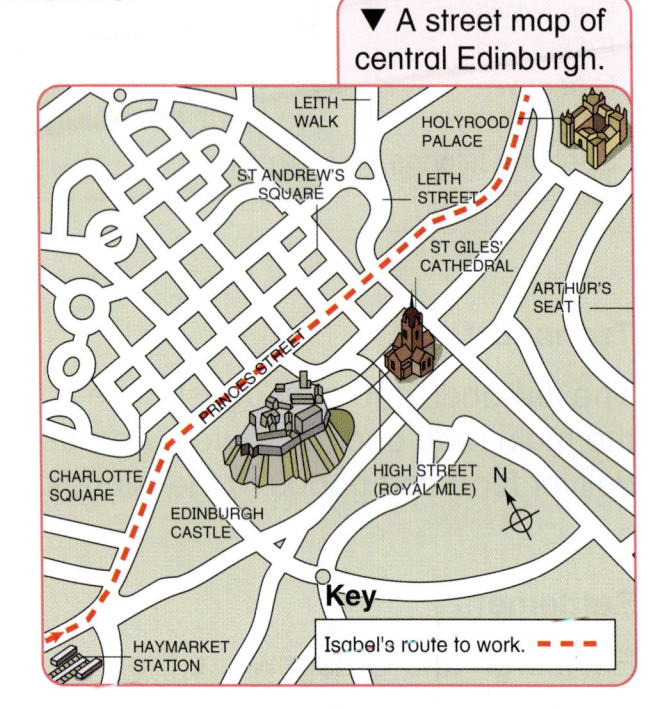

▼ A street map of central Edinburgh.

Key
Isabel's route to work. - - -

Discussion
- How can you tell Edinburgh has been an important city for a long time?
- How is Edinburgh different to the place where you live?
- Think of five places in Edinburgh where people might find work.

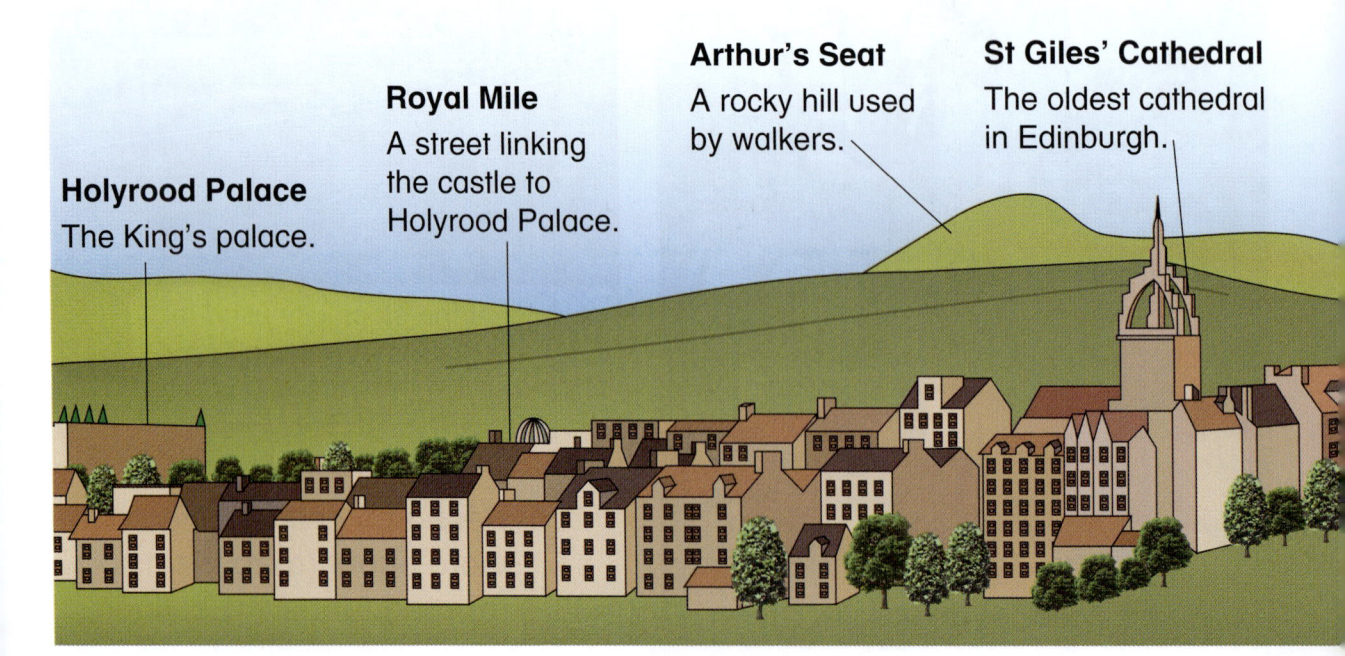

Holyrood Palace
The King's palace.

Royal Mile
A street linking the castle to Holyrood Palace.

Arthur's Seat
A rocky hill used by walkers.

St Giles' Cathedral
The oldest cathedral in Edinburgh.

Unit 7   Scotland

▲ A road and a rail bridge cross the Firth of Forth. The bridges are an important link between Edinburgh and other parts of Scotland.

Going to work

Isabel Andrews travels across Edinburgh every day on her way to work in a bank. The first thing Isabel passes is Haymarket station. Next she sees the castle. Isabel travels past the big shops on Princes Street. As she arrives at work she can see a rocky hill called Arthur's Seat. Millions of years ago this was a volcano.

Mapwork
Make a list of main landmarks which Isabel passes on her way to work.

Investigation
Research online to find out about three places you would like to visit in Edinburgh.

Key words
bank
castle
cathedral
crag
station

Edinburgh Castle
The castle used to protect the city from attacks.

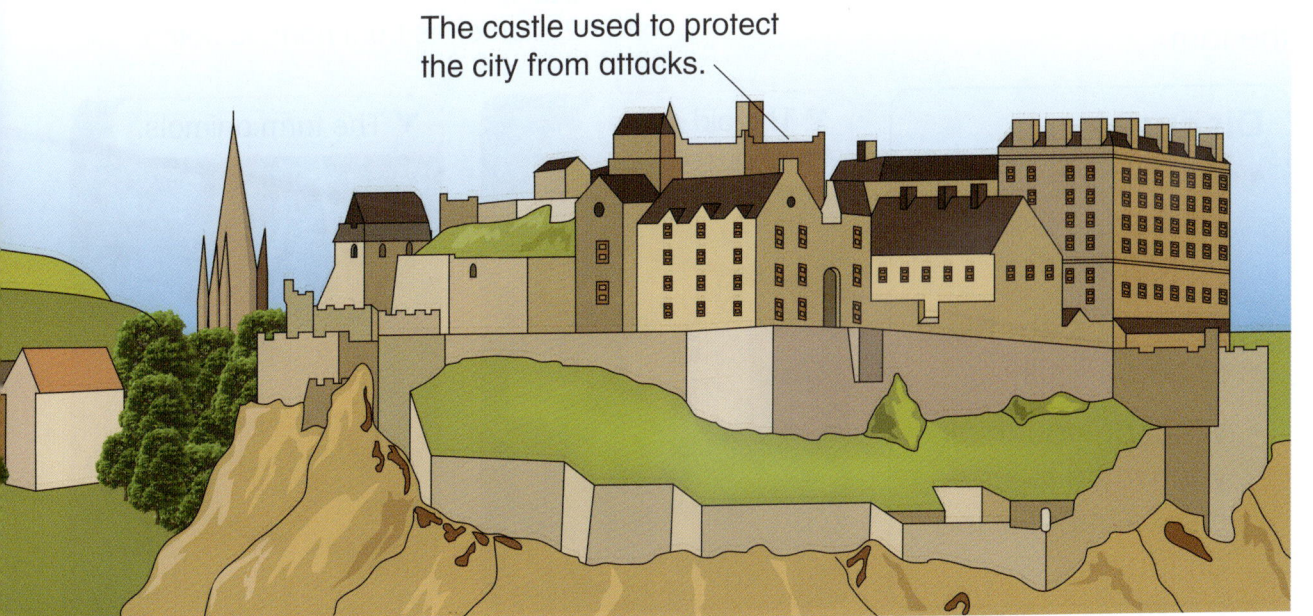

Unit 7 Scotland

Lesson 3: Mull: A Scottish island

▲ Colourful houses line the harbour at Tobermory, the main town on Mull.

Key words

cliff
croft
ferry
island
moor
tourist

Living on Mull

Iain and Morag live on the Isle of Mull. Mull is a very quiet island in Western Scotland. There are mountains and moors in the centre and cliffs along the shore.

Iain and Morag run a small farm called a croft. Iain looks after the sheep and cows on the farm. Morag knits beautiful jumpers. She sells them to tourists who come to stay on the farm.

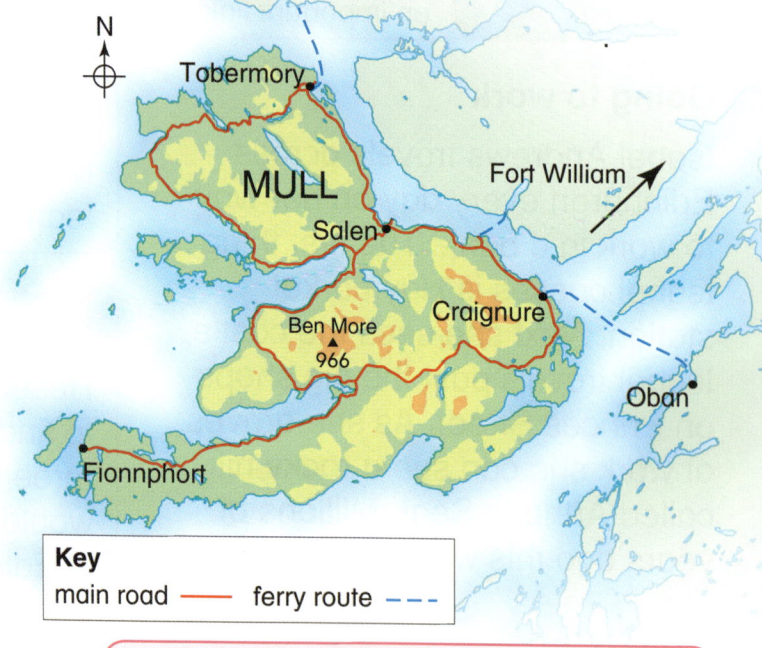

Key
main road —— ferry route ----

▲ Mull is 48 kilometres from west to east and 42 kilometres from north to south.

Discussion
- What is the landscape like on Mull?
- What jobs are there to do on Mull?
- What do you think you would like and dislike about living on Mull?

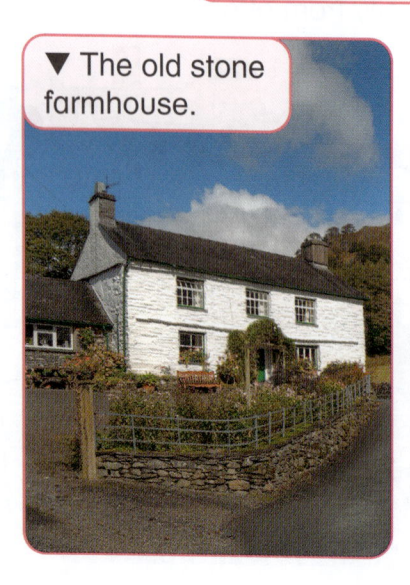

▼ The old stone farmhouse.

▼ The farm animals.

Unit 7 Scotland

Visiting Mull

Roy and Christine decided to go to Mull for their holidays because they enjoy bird watching. They arrived by ferry from Oban. The ferry was loaded with visitors, food and goods for the local people.

▼ View of Lismore lighthouse from Mull.

Christine bought a map so they could explore. Roy made sure they had waterproof anoraks in case it rained. They always took a picnic with them as there are very few shops outside Tobermory.

Investigation
Find out about the different things you can see and do in Western Scotland.

Data bank
- About 3000 people live on the Isle of Mull.
- There are four primary schools and one secondary school.
- Over 200 types of bird have been recorded on Mull.

◀ Here are some of the birds Roy and Christine saw on their walks.

Mapwork
Plan a holiday which would take you to three Scottish islands. Draw a map of your route.

Summary
In this unit you have learnt:
- what makes Scotland special
- about the capital city of Scotland
- about a Scottish island.

Unit 8 France

Lesson 1: Introducing France

What is France like?

France is about twice as large as the United Kingdom. It lies to the south of the United Kingdom across the English Channel.

▼ Skiing in the French Alps in winter.

Scale
0 100 200 300 km

Discussion
- How big is France?
- Using the photographs, what four things would you tell someone about France?
- What French things can be bought in your country?

Key
∧ Mountains

Key words
Alps
English Channel
Paris
River Seine

44

Unit 8 France

Landscape
The highest mountains in France are the Alps and Pyrenees. Their peaks are covered in snow and ice all year. The Seine, Loire, Garonne and Rhône are the longest rivers.

Weather
The north and west of France are often mild and quite wet. The south has hot, dry summers.

▲ Nôtre Dame and the River Seine, Paris.

Settlement
Paris is the capital city. Other large cities are Lyon, Marseille, Bordeaux, Lille and Strasbourg. There are many country areas with scattered villages.

Work
France has many industries such as iron and steel, glass and chemicals, cars and aeroplanes and making clothes. People also work on the land. They produce cheese, fruit, meat and vegetables.

▼ Many kinds of cheese are made in France.

▲ High speed trains can reach speeds of 300 km an hour.

Transport
France has one of the most modern transport systems in the world. There are high speed railway lines and motorways.

Mapwork
Using an atlas draw a map of ferry routes between France and the UK.

Investigation
Make a list of all the different ways your class or community is linked to France.

Unit 8　France

Lesson 2: Growing food

What crops do French farmers grow?

Parnac is a village in the Dordogne in the south-west of France. It has a church, old stone buildings and newer houses. The River Lot flows on one side of the village. Parnac has plenty of sunshine and the soil in the river valley is very good. Many crops are grown in the area.

Parnac

Grapes are the most important crop. They are picked and sold all over the world.

Some people keep geese and ducks. A type of meat paste, called pâté, is made from them.

Plums, apricots, apples and peaches grow in the orchards.

Tomatoes, beans and maize are grown for food.

46

Unit 8 — France

Key words
crops
export
settlement
soil
village

Discussion
- Why is Parnac a good place for a village settlement?
- What jobs do you think people do in Parnac?
- What do you think might make Parnac change?

Data bank
- There are at least 350 different types of French cheese (one for every day of the year).
- France exports more food than any other country in Europe.

There are lots of caves under the ground. Thousands of years ago people used to paint pictures on the walls. Today tourists go to see the famous paintings.

There are many fine old country houses called châteaux in the Dordogne.

The milk from goats is used to make cheese.

The nuts from walnut trees are used in cooking.

Mapwork
Use an atlas to find out approximately how far it is from London to Bordeaux (the city closest to Parnac).

Investigation
Make a list of the food that is grown at Parnac. Discuss if each one is also grown in your country. Show your answers with a tick.

Unit 8 France

Lesson 3: Making cars

Where do Renault cars come from?

Key words	
Europe	river
factory	sustainable
motorway	

Renault is an important car company. It has factories in France and other countries across Europe. One of the oldest is at Flins, a town about 40 kilometres from Paris.

Flins is near a motorway, railway and the River Seine. It is easy to reach by road and rail. There is also plenty of water. A lot of the people who work at the factory live in nearby towns and villages.

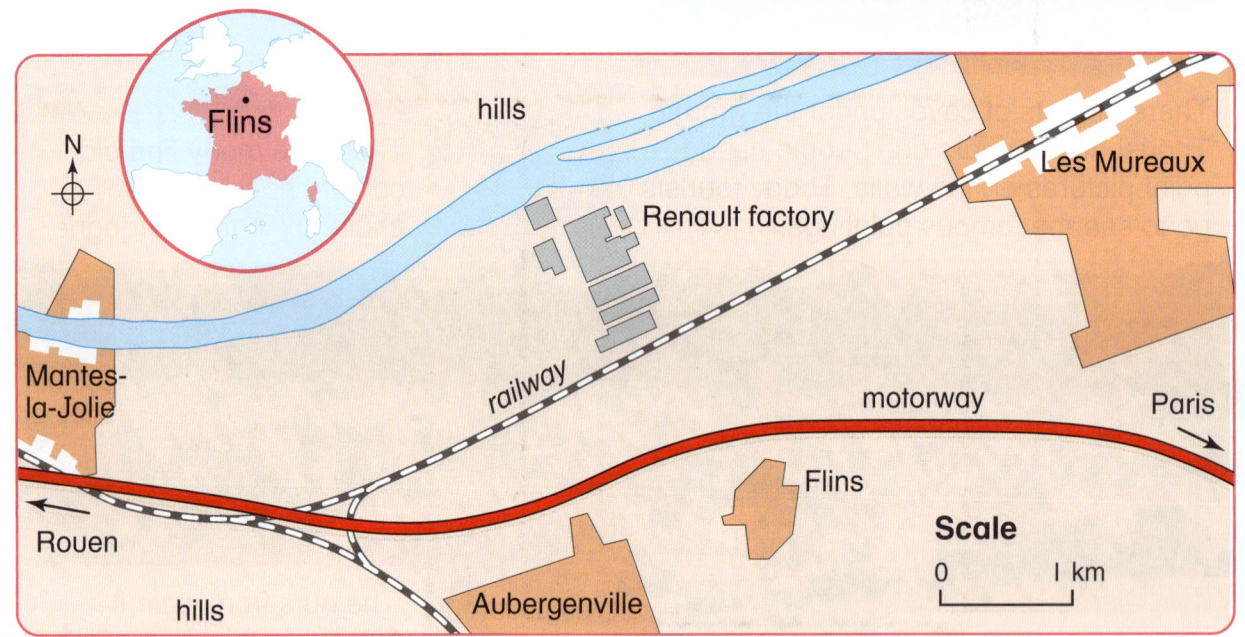

▼ The Flins car factory.

Data bank
- The Flins car factory opened in 1952.
- It has produced 18 million cars.
- Thousands of people work at the factory.

Discussion
- Why is Flins a good place for a car factory?
- What jobs do people do at Flins?
- What might happen to people in Flins if the factory closed?

Unit 8 France

Working in Flins

▲ Monsieur Hugo sprays the cars with paint.

▲ Madame Renard checks the cars are safe.

▲ Monsieur Hassan puts the engines into the cars.

▲ Madame Blanc runs the local supermarket.

Caring for the environment

Renault is setting up a new factory at Flins. This will make vehicles that use fewer resources and last longer. The factory will also reuse old parts such as batteries. This will be better for the environment, save energy and reduce waste. Other companies also want to stop what they make creating rubbish when they are worn out. They want to be more sustainable.

▼ Cars are now made of materials that last longer.

The production cycle

New cars.

Old cars are taken to pieces.

The parts are recycled.

Climate change

How do you think the new factory will be better for the environment?

Summary

In this unit you have learnt:
- about the landscape and weather of France
- about French food and crops
- about car-making in France.

Unit 9 South America

Lesson 1: Introducing South America

What is South America like?

South America stretches southwards from the Panama Canal to Antarctica. The Andes mountains run down the western edge of the continent. There are many snowy peaks and active volcanoes. The River Amazon rises here. It flows east for 6440 kilometres before it enters the Atlantic Ocean at the equator.

South America is divided into 13 countries. Brazil is much bigger than any of the others and covers half the continent. Five hundred years ago sailors from Spain and Portugal came to South America looking for gold and silver. Spanish and Portuguese are still the main languages in South America today.

▲ Bolivian woman on a mountain trail by Lake Titicaca.

Discussion
- What are the Andes?
- Why are European languages spoken in South America?
- What different things do the photographs tell you about South America?

▲ Rio de Janeiro, Brazil, is built around a bay and islands.

Unit 9   South America

Map

Labels on map: Panama Canal, River Orinoco, Angel Falls, Equator, River Amazon, PERU, Lima, Lake Titicaca, BOLIVIA, BRAZIL, Brasília, Rio de Janeiro, São Paulo, Pacific Ocean, River Paraguay, Mt Aconcagua, Buenos Aires, ARGENTINA, Atlantic Ocean

Scale: 0   1000   2000 km

Key
- Desert
- Grassland
- Forest
- Rainforest
- Mountain

South America

Key words
Andes
Brazil
equator
Lake Titicaca
Panama Canal
Rio de Janeiro
River Amazon

Data bank
- The River Amazon carries four times as much water as any other river in the world.
- Compared to other continents, more people live in cities in South America than in the countryside.
- Lake Titicaca is the highest lake in the world on which ships sail (3811 metres).

▼ The River Amazon flowing through the rainforest.

Mapwork
Make a list of countries in South America which are
(a) north of the equator
(b) south of the equator
(c) on the equator.

Investigation
Draw your own map of South America. Mark six places or features. Write a sentence about each one.

51

Unit 9 **South America**

Lesson 2: Spotlight on Chile

What is Chile like?

Chile is shaped like a long, thin ribbon. It stretches over 4000 km down the coast of South America but is only around 200 km wide. This makes it one of the most unusual countries in the world.

Chile is a country of contrasts. The Atacama Desert in the north is one of the driest places on Earth. In the central region around the capital, Santiago, there is good farmland. This is the most crowded part of Chile. In the south, forests and glaciers reach down to the sea. Here there are flooded valleys called fjords, similar to those in Norway.

Discussion
- What is special about the shape of Chile?
- What are the main regions in Chile?
- What three things do you think are most interesting about Chile?

Data bank
- Chile is about three times the size of the UK. When it is winter in the UK, it is summer in Chile.
- Santiago suffers from air pollution which becomes trapped by the Andes.
- There are geysers, hot springs and craters in the Atacama Desert.

Key words
copper
desert
fjord
geyser
glacier
hot spring
salmon
volcano

Key
- Mountain
- Desert
- Farmland
- Forest

Unit 9 South America

1 Copper mines

The Escondida mine in the Atacama Desert is the largest copper mine in the world.

3 Vineyards

Grapes grow well in central Chile. They are harvested and sent to supermarkets.

5 Salmon farms

The fjords in southern Chile are ideal for salmon. Chile is the world's biggest salmon producer after Norway.

Mapwork

Make simple outline drawings to show the shape of Chile, Norway and Vietnam.

2 Volcanoes

There are many volcanoes in the Andes. Lascar is one of the most active and has erupted many times.

4 Traditional crafts

Some people make traditional toys, gifts and clothes. Special deals ensure workers get a fair price for their goods.

Investigation

Make up a quiz. Write three sentences about Chile and three sentences about your country. Mix them up and see if someone else can guess the country.

Unit 9 South America

Lesson 3: The Galapagos Islands

Key words
- equator
- ocean current
- summit
- volcano
- heritage site

What is special about the Galapagos Islands?

The Galapagos are a group of remote islands about 1000 kilometres from South America. Over thousands of years, many different creatures have evolved in the Galapagos. Some of these cannot be found anywhere else in the world.

The Galapagos Islands were made into a World Heritage Site in 2001. Visitor numbers are strictly controlled. If new plants and creatures are brought from the mainland, they could upset the balance of life.

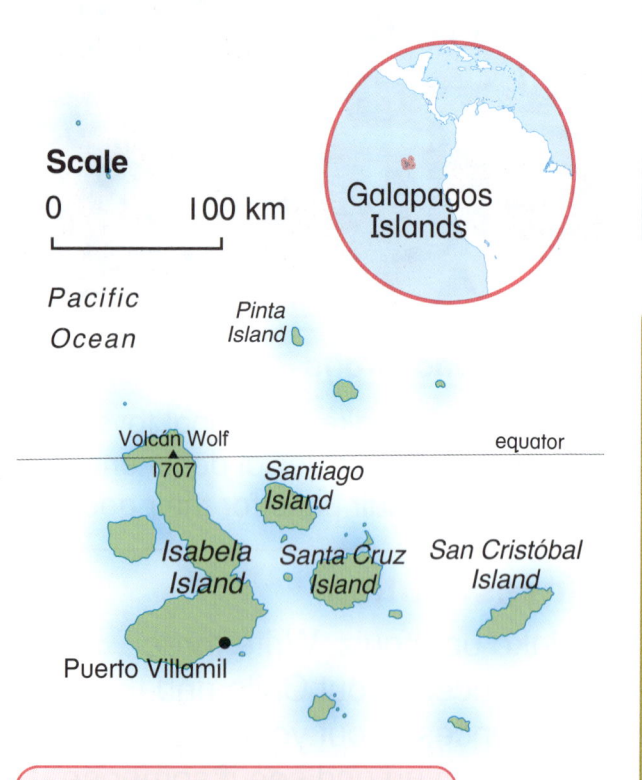

▼ The Galapagos Islands.

▲ There are over a hundred little islands in the Galapagos.

Discussion
- What makes the Galapagos Islands remote?
- What might upset the balance of life on the islands?
- Looking at the photographs, which plants and creatures do you think are most remarkable?

Data bank
- Three ocean currents meet at the Galapagos Islands, creating different ocean habitats.
- The Galapagos are the summit of an underwater volcano that rises 3000 metres from the ocean floor.
- 30 000 people live in the Galapagos Islands and there are 170 000 visitors each year.

Unit 9　South America

Charles Darwin

Charles Darwin visited the Galapagos Islands in 1835 when he was a young man. He spent many years thinking about the remarkable plants and creatures he found there, trying to explain why they had evolved. The theory that he came up with included new ideas about life on Earth.

▼ Sally lightfoot crab.

▼ Frigate bird.

▲ Giant tortoise.

▲ Land iguana.

Mapwork
Make a map of Isabela Island in your geography notebook. Write a sentence saying how far it is (a) across (b) round the coast.

Investigation
Make a small zigzag book about the Galapagos Islands and its wildlife.

Summary
In this unit you have learnt about:

- the features of South America
- what makes Chile special
- the plants and creatures of the Galapagos Islands.

Unit 10   Asia

Lesson 1: Introducing Asia

What is Asia like?

Asia is the largest continent. It is nearly 10 000 kilometres from east to west. In the north the climate is very cold. The south is hot with deserts and rainforests.

More than half the people in the world live in Asia. The largest countries are Russia, China and India.

Key words
China
Gobi Desert
Himalayas
India

Mapwork
Make a list of ten countries in Asia from an atlas.

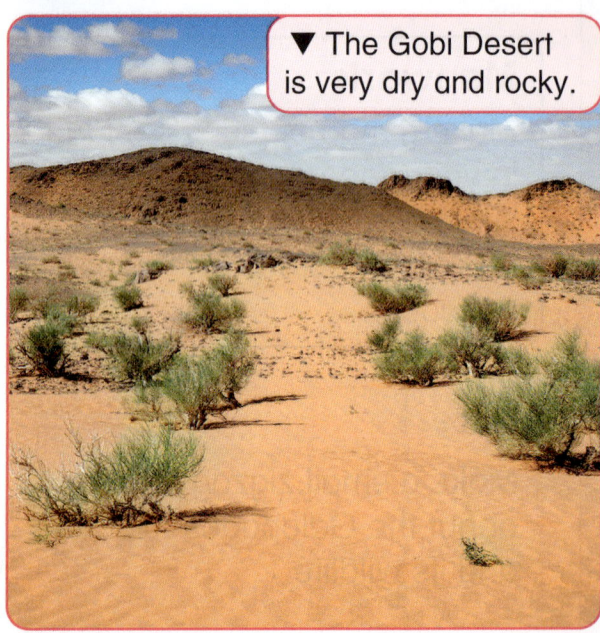

▼ Shanghai is one of the largest cities in China.

▲ The Himalayas are the highest mountains in the world.

▼ The Gobi Desert is very dry and rocky.

Unit 10 Asia

Asia

Key
- Mountain
- Desert
- Grassland
- Forest

EUROPE
Moscow
Siberia
RUSSIA
ASIA
steppes
Gobi Desert
Beijing
JAPAN
Tokyo
K2
CHINA
Shanghai
AFRICA
Arabian Desert
HIMALAYAS
New Delhi
Mount Everest
INDIA
Mumbai
Kolkata
Pacific Ocean
Indian Ocean

Scale
0 1000 2000 km

▼ The plains of Siberia are covered with pine trees and swamps.

▼ In Southeast Asia people depend on the monsoon rains to make their crops grow well.

Investigation

Copy the map key. Draw pictures to go with two of the landscape types.

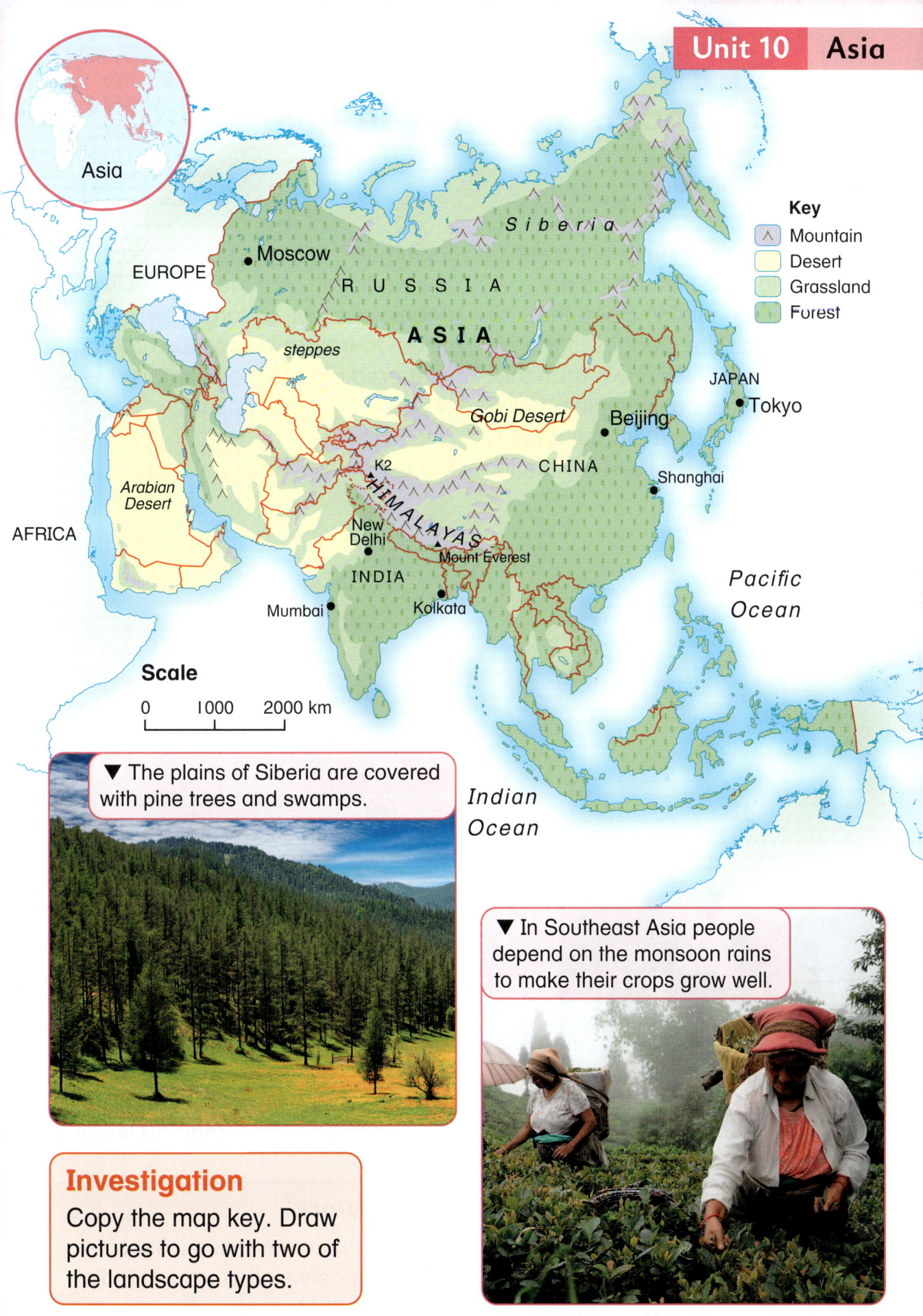

57

Unit 10 Asia

Lesson 2: India: A country in Asia

Key words
New Delhi
Himalayas
River Ganges
Thar Desert

What is India like?

India is the seventh largest country in the world. The Himalayas are a barrier of mountains along the north of India. The Ganges is the biggest river. It flows down from the mountains across a wide plain to the Bay of Bengal. There are deserts in the west and rocky hills in the south.

India has the largest population in the world with around 1500 million people. As well as large towns and cities, there are thousands of villages scattered across the countryside. New Delhi is the capital.

▲ Cricket is the national sport.

Investigation
Make a poster all about India. Include a map, photographs or drawings, as well as facts.

Mapwork
Using an atlas make a list of countries which surround India.

58

Unit 10   Asia

Lesson 3: Pallipadu: A village in India

What is life like in an Indian village?

Pallipadu is on the east coast of India. There are many quiet streets and plenty of trees in the village.

About 3000 people live in Pallipadu. Most of them are farmers. They grow rice, lentils, peanuts and vegetables. They also keep chickens and buffaloes.

▼ Rice field.

The soil is very fertile so people can grow two or three crops a year.

Between December and April the weather in Pallipadu is dry and warm. After that it becomes very hot and humid. When the monsoon rains come in July, it is cooler.

Once there was a cyclone which caused terrible floods. Over a hundred people were drowned.

◄ Farmers use buffaloes to plough the fields.

Key words

buffalo	monsoon
cyclone	rains
fertile soil	peanuts
flood bank	temple
lentils	wells

Jan	Feb	Mar	Apr	May	Jun	Jul	Aug	Sep	Oct	Nov	Dec
Pleasantly dry and warm			Hot and humid			Monsoon rains			Cyclones		

Unit 10 Asia

In the past all the water in Pallipadu used came from wells and pumps. Now there is a water tank and taps on the street. Having a safe supply of water is very important to the villagers.

Buildings are changing too. The old houses had thatched roofs and mud walls. Now most of the houses are made of brick and cement. This means they are not so easily knocked down by cyclones. However, the new houses are small and they get hot inside in summer.

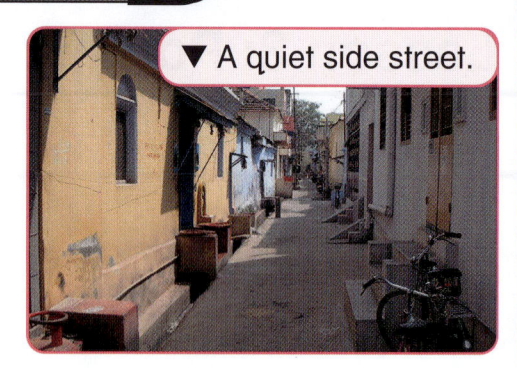

▼ A quiet side street.

Mapwork
Make a list and small drawings of the places shown on the map of Pallipadu.

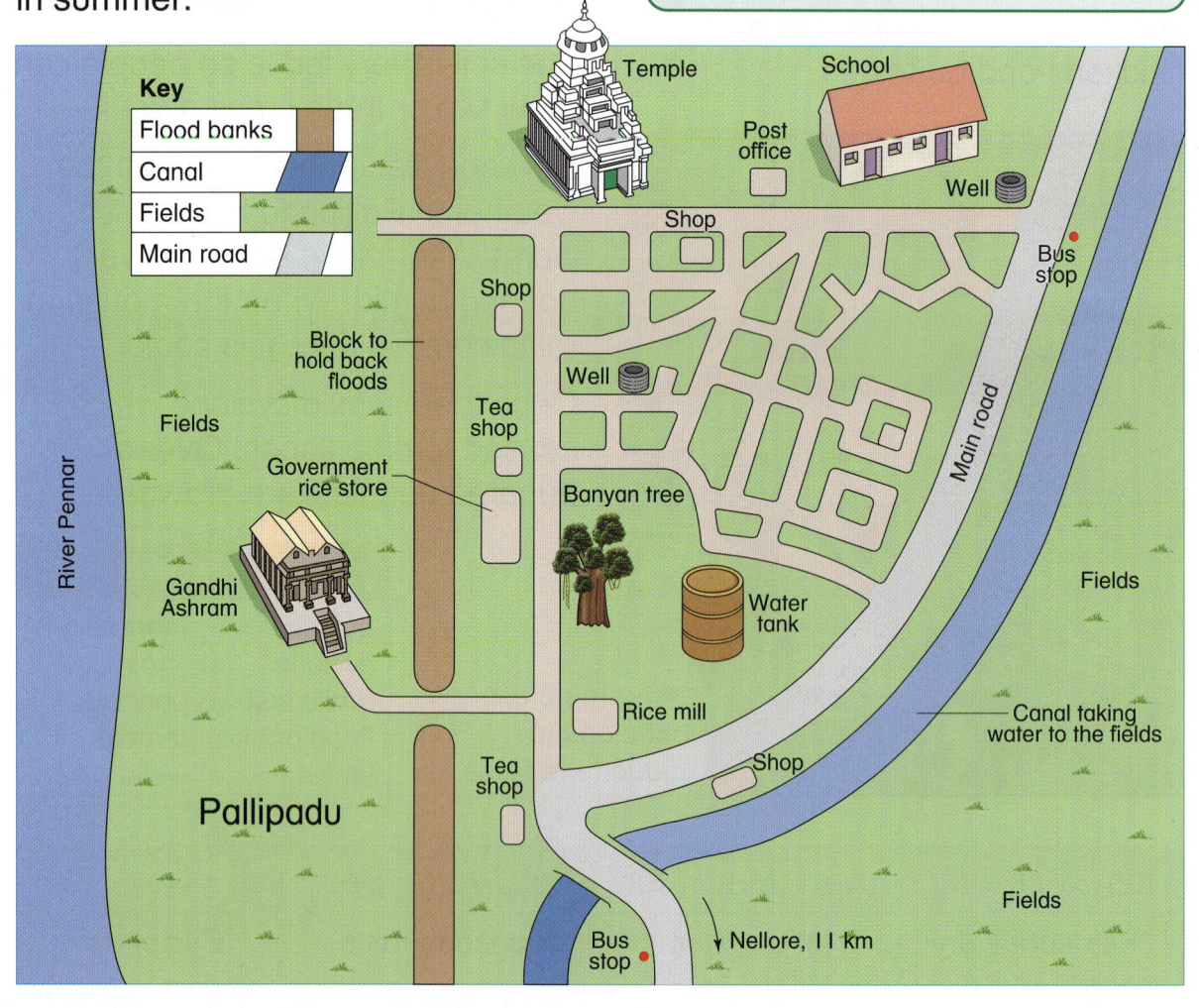

Key
- Flood banks
- Canal
- Fields
- Main road

Map labels: Temple, School, Post office, Well, Shop, Bus stop, Shop, Block to hold back floods, Well, Fields, Tea shop, Government rice store, Banyan tree, Main road, Gandhi Ashram, River Pennar, Water tank, Fields, Rice mill, Canal taking water to the fields, Shop, Tea shop, Pallipadu, Bus stop, Nellore, 11 km, Fields

Unit 10   Asia

Lots of people in Pallipadu use motor scooters and bicycles. There is a bus which goes to the local town. People go there to sell vegetables or to see a film. At different times of the year there are festivals. People pray in the temple and afterwards, there is dancing and everyone has a holiday.

▼ People enjoy special days like festivals and weddings.

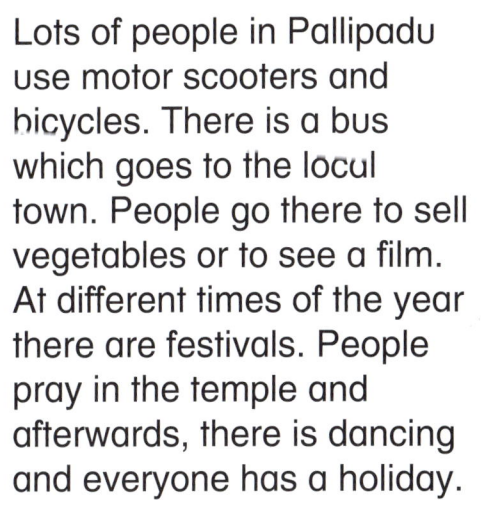

▼ There are many small shops in all parts of India.

Discussion
- Which are the best months to visit Pallipadu?
- What are people in Pallipadu doing about water security and cyclones?
- How is Pallipadu different to where you live?

Investigation
Imagine you are visiting Pallipadu. Write an email to a friend saying what you saw on a tour of the village.

Summary
In this unit you have learnt about:
- the landscape of Asia
- different parts of India
- life in an Indian village.

Glossary

Climate
The pattern of weather over many years.
Climate change
Long term changes in weather patterns.
Continent
Great blocks of land, such as Africa.
Cyclone
A fierce storm which affects parts of Asia.
Desert
A dry area where there is very little rain.
Environment
The world around us.
Equator
An imaginary line around the Earth, half-way between the North and South Poles.
Farm
A place where crops are grown and animals are kept for food.
Forestry
Growing trees so they can be sold for money.
Geography
The study of the surface of the earth and how people live.
Goods
The things which people sell to each other.
Habitat
The place where plants and animals live.
Hibernate
An animal hibernates when it goes to sleep during the winter months.
Landscape
The shape of the land, which can be made up of mountains, hills, valleys and other features.
Monsoon rains
Rainy weather which comes after the dry season in Southeast Asia.
North Pole
The most northerly point on the earth's surface.

Oasis
A place in the desert where there is enough water for trees and plants to grow.
Ocean
The seas which surround the continents.
Planet
A mass of rock and gas which circles around a star.
Plateau
A fairly flat piece of land high up in the mountains.
Polar lands
The land and ice around the north and south poles.
Pollution
Changes in our surroundings which damage the health of people, plants or animals.
Rainforest
Areas of thick forest which are near to the equator.
Settlement
The places where people live, such as villages, towns and cities.
South Pole
The most southerly point on the Earth's surface.
Sustainable
Using natural resources in a way that we can keep doing for a long time.
Tourist
Someone who visits places on holiday.
Transport
The vehicles used by people or goods.
United Kingdom
The country made up of England, Wales, Scotland and Northern Ireland.
Water vapour
An invisible gas in the air.